T0176012

How Artifacts Afford

Design Thinking, Design Theory

Ken Friedman and Erik Stolterman, editors

Design Things, A. Telier (Thomas Binder, Pelle Ehn, Giorgio De Michelis, Giulio Jacucci, Per Linde, and Ina Wagner), 2011

China's Design Revolution, Lorraine Justice, 2012

Adversarial Design, Carl DiSalvo, 2012

The Aesthetics of Imagination in Design, Mads Nygaard Folkmann, 2013

Linkography: Unfolding the Design Process, Gabriela Goldschmidt, 2014

Situated Design Methods, edited by Jesper Simonsen, Connie Svabo, Sara Malou Strandvad, Kristine Samson, Morten Hertzum, and Ole Erik Hansen, 2014

Taking [A]part: The Politics and Aesthetics of Participation in Experience-Centered Design, John McCarthy and Peter Wright, 2015

Design, When Everybody Designs: An Introduction to Design for Social Innovation, Ezio Manzini, 2015

Frame Innovation: Creating New Thinking by Design, Kees Dorst, 2015

Designing Publics, Christopher A. Le Dantec, 2016

Overcrowded: Designing Meaningful Products in a World Awash with Ideas, Roberto Verganti, 2016

FireSigns: A Semiotic Theory for Graphic Design, Steven Skaggs, 2017

Making Design Theory, Johan Redström, 2017

Critical Fabulations: Reworking the Methods and Margins of Design, Daniela Rosner, 2018

Designing with the Body: Somaesthetic Interaction Design, Kristina Höök, 2018

Discursive Design: Critical, Speculative, and Alternative Things, Bruce M. Tharp and Stephanie M. Tharp, 2018

Pretense Design: Surface over Substance, Per Mollerup, 2019

Being and the Screen: How the Digital Changes Perception, Stéphane Vial, 2019

How Artifacts Afford: The Power and Politics of Everyday Things, Jenny L. Davis, 2020

How Artifacts Afford

The Power and Politics of Everyday Things

Jenny L. Davis

The MIT Press
Cambridge, Massachusetts
London, England

© 2020 Massachusetts Institute of Technology

This book was set in ITC Stone Serif Std and ITC Stone Sans Std by New Best-set Typesetters Ltd. Printed and bound in the United States of America.

Library of Congress Cataloging-in-Publication Data

Names: Davis, Jenny L., author.
Title: How artifacts afford : the power and politics of everyday things / Jenny L. Davis.
Description: Cambridge, Massachusetts : The MIT Press, 2020. | Series: Design thinking, design theory | Includes bibliographical references and index.
Identifiers: LCCN 2019046429 | ISBN 9780262044110 (hardcover)
Subjects: LCSH: Manufactures--Psychological aspects. | Product design--Social aspects.
Classification: LCC TS171.4 .D395 2020 | DDC 658.5/752--dc23
LC record available at https://lccn.loc.gov/2019046429

10 9 8 7 6 5 4 3 2 1

This book is dedicated to my family: G, Hol, Rup, Mo, Nana, Carol, James, and The Boys.

Contents

Series Foreword

As professions go, design is relatively young. The practice of design predates professions. In fact, the practice of design—making things to serve a useful goal, making tools—predates the human race. Making tools is one of the attributes that made us human in the first place.

Design, in the most generic sense of the word, began over 2.5 million years ago when *Homo habilis* manufactured the first tools. Human beings were designing well before we began to walk upright. Four hundred thousand years ago, we began to manufacture spears. By forty thousand years ago, we had moved up to specialized tools.

Urban design and architecture came along ten thousand years ago in Mesopotamia. Interior architecture and furniture design probably emerged with them. It was another five thousand years before graphic design and typography got their start in Sumeria with the development of cuneiform. After that, things picked up speed.

All goods and services are designed. The urge to design—to consider a situation, imagine a better situation, and act to create that improved situation—goes back to our prehuman ancestors.

Making tools helped us to become what we are: design helped to make us human.

Today, the word *design* means many things. The common factor linking them is service, and designers are engaged in a service profession in which the results of their work meet human needs.

Design is first of all a process. The word *design* entered the English language in the 1500s as a verb, with the first written citation of the verb dated to the year 1548. *Merriam-Webster's Collegiate Dictionary* defines the verb *design* as "to conceive and plan out in the mind; to have as a specific purpose; to devise for a specific function or end." Related to these is the act of drawing, with an emphasis on the nature of the drawing as a plan or map, as well as "to draw plans for; to create, fashion, execute or construct according to plan."

Half a century later, the word began to be used as a noun, with the first cited use of the noun *design* occurring in 1588. *Merriam-Webster's* defines the noun as "a particular purpose held in view by an individual or group; deliberate, purposive planning; a mental project or scheme in which means to an end are laid down." Here, too, purpose and planning toward desired outcomes are central. Among these are "a preliminary sketch or outline showing the main features of something to be executed; an underlying scheme that governs functioning, developing or unfolding; a plan or protocol for carrying out or accomplishing something; the arrangement of elements or details in a product or work of art." Today, we design large, complex process, systems, and services, and we design organizations and structures to produce them. Design has changed considerably since our remote ancestors made the first stone tools.

At a highly abstract level, Herbert Simon's definition covers nearly all imaginable instances of design. To design, Simon writes, is to "[devise] courses of action aimed at changing existing situations into preferred ones" (Simon, *The Sciences of the Artificial*, 2nd ed., MIT Press, 1982, p. 129). Design, properly defined, is the entire process across the full range of domains required for any given outcome.

But the design process is always more than a general, abstract way of working. Design takes concrete form in the work of the service professions that meet human needs, a broad range of making and planning disciplines. These include industrial design, graphic design, textile design, furniture design, information design, process design, product design, interaction design, transportation design, educational design, systems design, urban design, design leadership, and design management, as well as architecture, engineering, information technology, and computer science.

These fields focus on different subjects and objects. They have distinct traditions, methods, and vocabularies, used and put into practice by distinct and often dissimilar professional groups. Although the traditions dividing these groups are distinct, common boundaries sometimes form a border. Where this happens, they serve as meeting points where common concerns build bridges. Today, ten challenges uniting the design professions form such a set of common concerns.

Three performance challenges, four substantive challenges, and three contextual challenges bind the design disciplines and professions together as a common field. The performance challenges arise because all design professions

1. act on the physical world,

2. address human needs, and

3. generate the built environment.

In the past, these common attributes were not sufficient to transcend the boundaries of tradition. Today, objective changes in the larger world give rise to four substantive challenges that are driving convergence in design practice and research. These substantive challenges are

1. increasingly ambiguous boundaries between artifacts, structure, and process;
2. increasingly large-scale social, economic, and industrial frames;
3. an increasingly complex environment of needs, requirements, and constraints; and
4. information content that often exceeds the value of physical substance.

These challenges require new frameworks of theory and research to address contemporary problem areas while solving specific cases and problems. In professional design practice, we often find that solving design problems requires interdisciplinary teams with a transdisciplinary focus. Fifty years ago, a sole practitioner and an assistant or two might have solved most design problems. Today, we need groups of people with skills across several disciplines and the additional skills that enable professionals to work with, listen to, and learn from each other as they solve problems.

Three contextual challenges define the nature of many design problems today. While many design problems function at a simpler level, these issues affect many of the major design problems that challenge us, and these challenges also affect simple design problems linked to complex social, mechanical, or technical systems. These issues are

1. a complex environment in which many projects or products cross the boundaries of several organizations, stakeholder, producer, and user groups;
2. projects or products that must meet the expectations of many organizations, stakeholders, producers, and users; and
3. demands at every level of production, distribution, reception, and control.

These ten challenges require a qualitatively different approach to professional design practice than was the case in earlier times. Past environments were simpler. They made simpler demands. Individual experience and personal development were sufficient for depth and substance in professional practice. While experience and development are still necessary, they are no longer sufficient. Most of today's design challenges require analytic and synthetic planning skills that cannot be developed through practice alone.

Professional design practice today involves advanced knowledge. This knowledge is not solely a higher level of professional practice. It is also a qualitatively different form of professional practice that emerges in response to the demands of the information society and the knowledge economy to which it gives rise.

In his essay "Why Design Education Must Change" (from *Core77*, November 26, 2010), Donald Norman challenges the premises and practices of the design profession. In the past, designers operated on the belief that talent and a willingness to jump into problems with both feet gives them an edge in solving problems. Norman writes:

> In the early days of industrial design, the work was primarily focused upon physical products. Today, however, designers work on

organizational structure and social problems, on interaction, service, and experience design. Many problems involve complex social and political issues. As a result, designers have become applied behavioral scientists, but they are woefully undereducated for the task. Designers often fail to understand the complexity of the issues and the depth of knowledge already known. They claim that fresh eyes can produce novel solutions, but then they wonder why these solutions are seldom implemented, or if implemented, why they fail. Fresh eyes can indeed produce insightful results, but the eyes must also be educated and knowledgeable. Designers often lack the requisite understanding. Design schools do not train students about these complex issues, about the interlocking complexities of human and social behavior, about the behavioral sciences, technology, and business. There is little or no training in science, the scientific method, and experimental design.

This is not industrial design in the sense of designing products, but industry-related design, design as thought and action for solving problems and imagining new futures. This MIT Press series of books emphasizes strategic design to create value through innovative products and services, and it emphasizes design as service through rigorous creativity, critical inquiry, and an ethics of respectful design. This rests on a sense of understanding, empathy, and appreciation for people, for nature, and for the world we shape through design. Our goal as editors is to develop a series of vital conversations that help designers and researchers to serve business, industry, and the public sector for positive social and economic outcomes.

We will present books that bring a new sense of inquiry to the design, helping to shape a more reflective and stable design discipline able to support a stronger profession grounded in empirical research, generative concepts, and the solid theory that gives rise to what W. Edwards Deming described as profound knowledge (Deming, *The New Economics for Industry, Government, Education*, MIT, Center for Advanced Engineering Study, 1993). For Deming, a physicist, engineer, and designer, profound knowledge comprised systems thinking and the understanding of processes embedded in systems, an understanding of variation and the tools we need to understand variation, a theory of knowledge, and a foundation in human psychology. This is the beginning of "deep design"—the union of deep practice with robust intellectual inquiry.

A series on design thinking and theory faces the same challenges that we face as a profession. On one level, design is a general human process that we use to understand and to shape our world. Nevertheless, we cannot address this process or the world in its general, abstract form. Rather, we meet the challenges of design in specific challenges, addressing problems or ideas in a situated context. The challenges we face as designers today are as diverse as the problems clients bring us. We are involved in design for economic anchors, economic continuity, and economic growth. We design for urban needs and rural needs, for social development and creative communities. We are involved with environmental sustainability and economic policy, agriculture competitive crafts for export, competitive products and brands for micro-enterprises, developing new products for bottom-of-pyramid markets and redeveloping old products for mature or wealthy markets. Within the framework of design, we are also challenged to design for extreme situations; for biotech, nanotech, and new materials; for social business; as well as for conceptual challenges for worlds that do not yet exist (such as the world beyond the Kurzweil singularity) and for new visions of the world that does exist.

The Design Thinking, Design Theory series from the MIT Press will explore these issues and more—meeting them, examining them, and helping designers to address them.

Join us in this journey.

Ken Friedman
Erik Stolterman

Editors, Design Thinking, Design Theory Series

Acknowledgments

This book began with a student's curiosity, became viable through a generous act from a senior scholar, and materialized with the help of an incredible support team made up of colleagues, friends, and family—many who occupy more than one of these categories.

In 2015, I walked into a classroom at James Madison University in Harrisonburg, Virginia, prepared to lecture about technological affordances for a small undergraduate course. I did not expect to start a book. However, a young man named Ben asked just the right question ("Don't rope and wood fences afford differently?"), which led to writing on the classroom whiteboard, a series of blog posts,[1] a journal article in *Bulletin of Science, Technology & Society*,[2] and now, a full monograph. Ben's thoughtfulness was, literally, inspirational. In 2017, in a new appointment at the Australian National University (ANU) in Canberra, I booked a meeting with Professor Genevieve Bell, a recent transplant from Silicon Valley and founder ANU's Autonomy, Agency, and Assurance Institute (3Ai).[3] Within a day of that meeting, I had an introduction email to Doug Sery, acquisitions editor for the MIT Press. It was a small act for Professor Bell but monumental for me. Doug was interested in the project and ushered it through from start to finish, with tireless help from Noah J. Springer, whose dissertation, it turns out, addressed affordances (the topic of this book).[4]

I am forever grateful to D'Lane Compton, who shared her wisdom, time, and materials when I presented my full naivety in the form of a one-line Twitter message that read "how does one write a book?" I was further aided in both process and content through extensive conversations with and feedback from David A. Banks, Nathan Jurgenson, and PJ Patella-Rey. Their careful balance between encouragement and challenge kept my spirits up and my framing sharp. I am also grateful to all past and present contributors to the *Cyborgology* blog[5] and Theorizing the Web conference,[6] who have cultivated a community of thinkers, writers, and activists doing the kind of theory I want to see in the world. Equally influential and generous have been my closest collaborators on other projects. Tony Love, Carla Goar, and Bianca Manago have picked up my slack while I worked on the book, and each has made me a better thinker through years of intellectual discussion, debate, and endless Track Changes. The book was also helped along by three emerging scholars who gave me the privilege of supervising their honors theses. Siobhan Dodds, Hannah Gregory, and Will Orr asked about the book each week during our meetings, motivating me to write. They also kept me on my intellectual toes, the benefits of which cannot be overstated. Finally, thank goodness for James Chouinard. My coauthor, editor, and partner. Not once did you hesitate when I asked, "Is it okay if I read something out loud?" You have cheered me through this book while making sure that the ideas are sound. With you, I have full support and get away with nothing. Truly, thank you.

1   Introduction

A Trolley Problem of a Particular Sort

In January 2017, I relocated from the United States to begin an academic appointment in Canberra, Australia. This moment was marked by competing pulls of excitement and trepidation. The allure of adventure and the esteem I felt for my new institution were punctuated by anxiety about the unknown and uncertainty about life abroad. I had been warned that Australia was unlike America, despite the familiarity of a shared language. Heeding this advice, I spent my first weeks in Canberra watching others with anthropological vigilance, certain I would order coffee incorrectly or breach public transit decorum. I kept my voice at a soft timbre and Googled everything before I did it. I was determined to blend in, which I did successfully, for a while.

My first fish-out-of-water moment came unexpectedly, and it had nothing to do with Australian culture. In fact, it was tied to an activity for which I had presumed full competence: acquiring a shopping cart or, in Australian parlance, a shopping trolley. It was a hot day in the peak of summer and I was moving from temporary campus housing to a more permanent place outside the city. Having left behind nearly all my worldly possessions,

I needed starter supplies to set up a new home. After a quick internet search for "how to get gas in Australia" and a precarious drive on the left side of the road to a nearby big-box store, I took a deep breath and looked for the largest shopping cart I could find.

To my surprise, I found only hand-held baskets and carts that were linked and locked together. I asked a clerk, "Do you have any trollies available for immediate customer use, and if not, could you please unlock one for me?" The clerk informed me that the trollies took a $2 coin deposit. Besides the fact that I had no idea Australia's currency included $2 coins, I verged bewildered: "Are you telling me I need to pay to use a cart?" The clerk blinked, started to explain, and then used a key around his belt to unlock a cart before sending me on my way.

After a few moments of studying the cart's blue handle—it had three small currency slots, a lock device, and an opening into which the lock device fits—I understood. Customers don't rent the carts, but use coins as collateral. When returning the cart, shoppers retrieve their money by locking the used cart back in place, which releases the coin deposit.

Coin-locks are a theft-prevention measure and a now common feature of commerce in many urban environments. However, because I grew up in the suburbs and lived in small towns for most of my adult life, coin-locks were new to me. I was used to seeing shopping carts that were free-standing and abundant. In fact, I once lived in an apartment complex in Texas with an informal shopping cart repository in the parking lot. The local supermarket chain sent employees to retrieve the carts once a day. But in Australia's capital city, coin-locks are standard.[1]

The problem of shopping cart retention is an ironic one in the context of the cart's history. In 1937, Sylvan Goldman introduced the wheeled shopping cart to reluctant customers at his Humpty Dumpty grocery chain in Oklahoma. By that time, the design of shops had shifted from a model where clerks stood behind a counter and fetched items for customers to a self-service model where customers selected their own items from displays around the store.[2] At first, customers used hand-held baskets to collect and deliver their goods to the checkout counter. As store sizes expanded and grocery loads grew, the conventional hand-held baskets proved less convenient. Clerks had to watch for customers with full baskets, hold customers' items until checkout, and provide fresh baskets for continued shopping. This could be inconvenient for shoppers and relied on paid labor from store staff. Goldman's wheeled cart model—which looks similar to the carts used in most stores today—enabled shoppers to buy more goods with greater convenience, while undercutting staffing costs.

Goldman's customers needed convincing. Women rejected the idea of pushing a cart because it too closely resembled a baby buggy. Apparently, women wanted shopping to feel like a break from childcare, not an extension of it. Men found carts too effeminate and rejected them on normative gender grounds. So Goldman mobilized a public relations and outreach campaign. Along with advertisements, Goldman hired attractive men and women to use shopping carts in his stores. The tactic worked. Shopping carts quickly spread to other retail outlets, becoming a fixture in the contemporary marketplace.

If Goldman had trouble persuading people to adopt his new technology, the existence of coin-locks represents an opposite problem: persuading people to give back the carts they've taken. The coin-lock was patented in various forms during the 1980s and 1990s and is one of several theft-prevention measures. Others

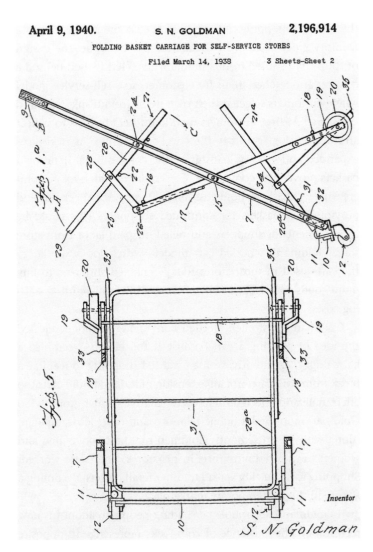

Image of Sylvan Goldman's early shopping cart

include electronic and magnetic features that lock a cart's wheels when it passes a perimeter; long poles attached to shopping carts that block them from fitting through exits; global positioning system (GPS) trackers; and even services that find, retrieve, and return carts for a fee. Not only do stolen or misplaced carts place a financial burden on stores (which pay from $150 to $400 for each replacement), but cities struggle with safety issues when stolen carts are left in roads, on sidewalks, and in creeks and streams. In short, both shops and cities have an interest in keeping shopping carts on company property, and developments in theft-prevention technologies reflect these interests. For customers, theft-prevention features may be a mere inconvenience (they need to remember to carry change) or may dramatically affect the flows of daily life (people without vehicles cannot easily transport large purchases by foot and so must allot time each day to stop by the store and buy provisions).

The evolution of the shopping cart from a labor-replacing technology that encourages high-volume purchases to a tightly controlled commodity fitted with material constraints shows that objects, even the most mundane, are imbued with values that reflect and have the capacity to shape social, political, and economic relations. Goldman's initial shopping cart was created under the drive of capital accumulation. The cart maximized buying while minimizing paid human labor. Cart usage (or lack thereof) was linked with issues of gender: women wanted to distance the shopping experience from the work of childcare, and men wanted to distance themselves from effeminate connotations of womanhood. Commercial strategies paved the way for widespread shopping cart adoption, and eventually, some carts and shops were redesigned in ways that limited and regulated cart use, with varying effects on consumers.[3] In short, the

shopping cart has politics, affects behavior, and shapes the flow
of daily life. These dynamics are built into the cart's material
form, with results that are subtle, powerful, and far reaching.

Affordances

This book is about the social dynamics of technology. It is about
the ways that ethics, values, and interests are built into technological objects and the ways these objects take shape through
interactions with human subjects. More specifically, this book
is about technological affordances. Formally, an affordance
is defined as "the 'multifaceted relational structure'[4] between
an object/technology and the use that enables or constrains
potential behavioral outcomes in a particular context."[5] That is,
affordances mediate between a technology's features and its outcomes. Technologies don't *make* people do things but instead,
push, pull, enable, and constrain. Affordances are *how* objects
shape action for socially situated subjects.

The concept of affordance was first introduced by the ecological psychologist James J. Gibson in the 1960s and 1970s.[6] For
Gibson, "affordance" was a way to approach the mutual constitution between people and environments. Donald A. Norman
brought affordances to design studies a decade later to address
human-machine interactions.[7] In recent years, the concept has
picked up considerable steam as the study of computer-mediated
communication (CMC) and information communication technologies (ICTs) have become firmly entrenched in the academic
canon.

It is unsurprising that the concept of affordance has surged
amid vast and rapid technological change. The ubiquity of smartphones, infusion of digital platforms, and rise of automation are

(re)shaping social relationships, information flows, political participation, and economic relations. Social thinkers are eager to understand these societal shifts and are thus interested in how new technologies work and to what effect. "Affordance" is a useful conceptual tool in such a project because it lets analysts interrogate the effects of emergent technologies while avoiding hardline determinism.

Technology studies offers the persistent reminder that materiality and human agency always operate together. Hence, Goldman's shopping cart does not force customers to purchase more goods, and hand-held baskets do not stop customers from buying in bulk. However, carts and baskets have features that differ in ways that structure the shopping experience and alter the distribution of labor between employees and consumers. In this way, front-facing digital cameras don't make people to take selfies but afford this photographic convention in ways that diverge from the affordances of traditional film-reliant devices. Touch-activated dictionaries interact with vocabulary acquisition differently than paper-bound volumes do. Drop-down menus shape choice in more confining ways than write-in boxes do. And large "REPORT" buttons on social media platforms afford user-generated content moderation differently than an administrator email hidden behind several clicks.

The analytic balance between materiality and human agency makes affordance a valuable concept that has sustained over time and spread across disciplines. However, the scholarly application of affordance has outpaced its careful theoretical consideration. The concept has been mired by misuse, overuse, false binaries, and inadequate treatment of dynamic subjects and circumstances. For these reasons, some scholars argue that the concept has lost analytic value and should be relinquished altogether.[8] As evidenced by my book-length attention to the topic, I believe this response is wrongheaded. Instead, I read the critiques of affordance as an opportunity for clarity and precision, and the concept's ascent alongside technological advancements as an indicator that such clarity and precision are needed now more than ever.

One persistent critique is that affordance has remained a binary construct. In its binary depiction, features either afford some action or do not afford that action. Coin-locked carts either afford transportability or do not; social media platforms either afford network building or do not; artificial intelligence (AI) either affords emotional attachment or does not. By this logic, features make actions either inevitable or impossible. In practice, we know that the relationship between people and things is never cut and dry. Human-technology relations are a subtle dance in which technological objects push and pull with varying degrees of insistence while human subjects navigate with more and less motivation, creativity, and skill. Concretely, the coin-lock system does not unequivocally or universally preclude the removal of shopping carts from store premises but instead creates conditions that make removal less likely. Indeed, while researching the history of the shopping cart, I found many tutorials and products aimed at surpassing wheel-locks, coin-locks, and GPS tracking devices. Thus, affordances are never determinations, nor are they uniform. Instead, features apply varying levels of pressure on socially situated subjects.

Luckily, affordance's binary problem has a simple analytic solution: shifting from questions about *what* technologies afford to *how* they afford. The shift from what to how undergirds the argument I delineate throughout this text. As a general rule, social analyses are much richer when approached with questions of how rather than what. The how captures processes and nuances, while the what remains one dimensional. By asking how technologies afford, we can identify and articulate variation in a given feature's social impact.

For instance, compared to systems without theft-prevention instruments, the coin-lock system creates a light barrier to using shopping carts. To use a coin-locked cart, customers need the proper resources (usually coins of a particular currency) and need to be willing to engage in extra tasks to obtain the cart at both the front and back ends of a shopping trip. These tasks include finding a coin, unlocking a cart, returning the cart when they are finished, locking it, and retrieving their coin. In practice, these actions take only about thirty extra seconds of work and are relatively inconsequential for many people in most circumstances. Nevertheless, the coin-lock feature creates friction and is thus antithetical to Goldman's early initiative to make carts as appealing and available as possible. The coin-lock prevents people from stealing the carts but also dissuades them from using the carts at all. Such dissuasion, however, is milder than if carts were held behind a counter and dispensed only by a clerk or if carts kept their coin deposits, which would change the system from collateral to rent-based (though the latter would also disincentivize cart return).

In comparing features of different theft prevention implements, both a coin-lock apparatus and magnetically triggered wheel-locks reduce the transportability of grocery carts, but the coin-lock system generally presents fewer barriers to taking carts off-site. A customer who takes a coin-locked cart off-site may lose $2, but the wheel-locked trolley stops rolling after crossing a perimeter. Both coin-locks and wheel-locks reduce transportability, but they do so with varying degrees of force, and neither makes the cart entirely nontransportable. Customers who encounter coin-locks may elect to forgo their $2 investment, leave the store with the cart and then come back to recoup their $2, use a universal cart key (they are easily found and purchased online), or simply wait to find a loose cart and take that cart off the lot. Customers who encounter wheel-locks may lift the cart over the magnetic locking strip, push the cart over the magnetic perimeter with significant force, load the cart into a vehicle, or if especially motivated and sufficiently able, carry the cart after the wheels go into lock mode. The point is that asking *how* instead of *what* objects afford shows nuanced relationships between technical features and their effects on human subjects while accounting for creative and subversive human acts.

A second critique is that analysts too often depict affordances as universal when in fact, they are relational and conditional. Given that technical features exert varying degrees of force, the next question to ask is *for whom and under what circumstances?*. For example, for me as a coin-lock novice, the coin-locks posed a stronger barrier to use than they would for customers more familiar with the system. Over time, I became accustomed to Canberra's coin-locks, and the affordances varied between my past and present selves. The barrier to use amplifies when I'm in a hurry (am I willing to expend the extra thirty seconds?) and reduces when I'm not on a schedule. The consequences for taking a shopping cart off site are relatively minor for me ($2 will not noticeably affect my bank account), but may be more consequential for someone experiencing homelessness or fending off hunger. (The need to take a cart off site may also be more pronounced for people in the latter group, who are less likely to have personal transportation and may use the cart for reasons other than grocery shopping).

In short, affordances refer to *how* objects enable and constrain. This will vary across people and contexts. Shifting from *what* to *how* and accounting for diverse subjects and circumstances represent a simple but crucial advancement in affordance theory. A more substantial advancement, which is the main project of this book, is to operationalize the concept of affordance such that *how, for whom,* and *under what circumstances* are incorporated into a concise analytic tool.

Operationalizing Affordances: The Mechanisms and Conditions Framework

This book delineates the *mechanisms and conditions framework* as a theoretical scaffold for affordance analyses. The mechanisms of affordance refer to the how of human-technology relations, and the conditions refer to variability across subjects and circumstances. Rather than rely on general statements about more and less force exerted by technological objects, the mechanisms of affordance indicate that technologies *request, demand, encourage, discourage, refuse,* and *allow* particular lines of action and social dynamics. *Requests* and *demands* are initiated by the object, and *encouragement, discouragement,* and *refusal* are responses to subjects' inclinations. *Allow* applies to acts initiated by both subjects and objects.

The conditions of affordance specify the relational nature of human-technology encounters—namely, the conditions of affordance vary by *perception, dexterity,* and *cultural and institutional legitimacy.* That is, people perceive a range of functions and constraints presented by technological objects, have varying

levels of skill in operating a set of features, and experience dif-
ferential support in engaging with a technology due to cultural
norms and institutional regulations.[9]

Operationalizing affordances through the mechanisms and
conditions framework provides a vocabulary and structure with
which to approach affordance analyses. For example, with the
mechanisms of affordance, we may say that shopping carts
encourage large purchases and hand-held baskets *discourage* large
purchases. In this vein, the hand-held baskets *request* frequent
trips to the shop, and the carts *encourage* fewer trips. Neither
baskets nor carts *refuse* frequent or infrequent shopping trips,
but they nudge shoppers in one direction or the other. Shoppers
using baskets and carts are *allowed* to fill their shopping vessels
with sale items, specialty items, frozen goods, or fresh produce
(that is, baskets and carts pay no mind to their contents outside
of weight and dimensions).

The conditions of affordance let us further parse the push and
pull of technologies by their circumstances of use. For example,
in the 1930s, the perceived link between carts and baby car-
riages *discouraged* use by women and men—who experienced the
apparatus as an extension of care labor and prohibitively femi-
nine, respectively. Goldman's early public relations campaign
was aimed at rebranding the shopping cart as a gender-neutral
labor-saving tool, thus *encouraging* use by shoppers across gender categories (and in turn, *requesting* that shoppers purchase
more goods in a single trip). Notably, despite Goldman's successful efforts to change perceptions and cultural norms, the
traditional cart model still *refuses* use by portions of the population. For example, those who use wheelchairs may not have the
physical dexterity to utilize Goldman's original cart design. The
cart therefore *encourages* use by walking customers but *refuses* use

among those with certain mobility impairments. Subsequent cart designs that include an adult-sized seat and motorized components undo this *refusal* and instead *encourage* adoption by those for whom walking is difficult or impossible.

Theft-prevention features also work differently depending on context. Wheel-locks *refuse* transportability for people who perceive no workarounds to magnetic perimeters but merely *discourage* transportability for those who are aware of alternatives (such as lifting the cart over the magnetic strip or pushing the cart with enough force to beat the lock device). Similarly, coin-locks *request* that users keep the carts on store premises but *allow* people with the requisite resources to move carts beyond store boundaries. Normative and implicit biases also apply here, as cultural and demographic markers can either mitigate or amplify surveillance, highlighting the relational dynamics of affordances in practice. For instance, customers of color are more likely to be followed by a store employee as they shop, thus *refusing* cart removal in a way that is merely *discouraged* for white customers, whom employees are more likely to grant freedom of movement around the store.

In short, technologies are efficacious in ways that manifest variously across persons and circumstances. The mechanisms and conditions framework offers a conceptual scaffold with which to address these dynamics. The mechanisms of affordance specify *how* technologies afford, while the conditions of affordance situate technologies in context. Crucially, the mechanisms and conditions framework is not a reifying device, but a tool of argumentation. The mechanisms of affordance are neither rigid nor determinative. Rather, they are analytic stopping points with porous boundaries, and the designation of one category versus another remains always up for debate. In turn, the conditions of affordance are neither static nor mutually exclusive but overlapping and always subject to change. The mechanisms and conditions framework thus provides a schematic onto which analysts and practitioners can map sociotechnical systems, maintaining the richness of dynamism, uncertainty, and robust deliberation.

How Affordances Matter

The mechanisms and conditions framework is rooted in the assumption that technologies are political. I address this base assumption more thoroughly in chapter 3. For now, I use the politics of technology to make a case for how affordances matter. Technologies are designed, implemented, and used through webs of choices. Some of these choices are explicit and reflect a clear intention for the technology to affect human action in some specific way. Other choices are implicit and may not ever enter the conscious minds of designers, distributors, or end users. Each choice—explicit or implicit—reflects and affects value orientations, sociostructural arrangements, and social dynamics.

Because values are not neutral and tend to reinforce power and status structures, technologies are often infused with the politics of the powerful. This is not to say that technologies cannot effect change for oppressed groups or serve as tools of resistance. They can, and they do. However, the mechanisms and conditions framework begins with the assumption that *if left unchecked*, technologies will arc toward privilege and normality. This assumption bears out empirically and repetitively. For example, several versions of facial recognition software have failed to identify dark-hued skin tones, thus excluding people of color from available services while reentrenching default whiteness; Facebook's real-name policy proved exclusionary and at times dangerous for some LGBTQI users; and a study by Carnegie Mellon University showed that Google's automated targeted ad feature presents men with higher-paying employment opportunities than those presented to women.[10]

The politics of technology stem from objects' integration with human social and structural arrangements. By asking *how*, *for whom*, and *under what circumstances?*, the mechanisms and conditions framework takes a relational position in which humans and technologies are inherently co-constitutive. Although technologies maintain a shaping effect on human subjects, technologies themselves embody human values and politics in their design, implementation, and use. The bad news is that this means technologies will, by default, reflect and reinforce existing inequalities. The good news is that the default is neither necessary nor inevitable. A sharp analytic tool, like the mechanisms and conditions framework, renders politics visible and pliable. Inclined practitioners can thus rework sociotechnical systems toward social good.

Situating the Text

A substantial body of work focuses on the entwinement of social and technical systems. This has emerged as a robust and interdisciplinary approach to the politics and values of technologies in society. From social science, we see rigorous analyses that detail the ways in which technical systems reflect and perpetuate inequalities along intersecting lines of race, class, sexuality, (dis)ability, geography, and gender. From engineering and design studies, we see an effort to integrate values, ethics, and politics into design processes. A properly operationalized model of affordances connects these intellectual and practical efforts by giving language and structure to projects that map the social dynamics of technical systems and to projects that design technical systems with social intent.

Virginia Eubanks's *Automating Inequality: How High-Tech Tools Profile, Police, and Punish the Poor*[11] and Safiya Umoja Noble's *Algorithms of Oppression: How Search Engines Reinforce Racism*[12] stand out as exemplar works from the social sciences. In design studies, there has been a "practical turn" exemplified by Batya Friedman and colleagues' work on value-sensitive design[13] and Mary Flanagan and Helen Nissenbaum's research on *Values at Play in Digital Games*.[14] I highlight these works here to situate the mechanisms and conditions framework of affordances within a larger cross-disciplinary project of critical approaches to technology and design. I also highlight them to show the utility of the mechanisms and conditions framework as a cohesive analytic and practical tool.

Eubanks's *Automating Inequality* documents the effects of automated decision systems in the US public sector. Billed by government agents as objective and optimally efficient, automated systems have been mobilized to manage public welfare, healthcare, homelessness, and children's protective services. Eubanks shows that as they are built, these automated systems over-monitor and underserve populations in need. For example, any missing data for a user in the healthcare distribution system resulted in an immediate cease of benefits with no clear information about what the problem was or how to fix it. Recipients would simply receive notification that they were unable to access benefits, and the burden was placed on the beneficiary to reconcile with the system. People experiencing homelessness were required to answer a battery of questions to be eligible for housing, thus placing them in databases for surveillance and monitoring by police and government authorities (while remaining highly unlikely to receive sustainable housing assistance). Automated systems for child protection relied on a point-based algorithm that predicted the likelihood that a child would experience danger. The algorithm was predicated largely on interactions between the family and public services, thus placing poor families under disproportionate scrutiny and increasing the likelihood that parental custody would come under threat. In short, Eubanks shows that "poor and working-class people are targeted by new tools of digital poverty management and face life-threatening consequences as a result."[15]

Noble's *Algorithms of Oppression* examines algorithmic biases at the intersection of race and gender in the Google search engine. Opening with an account of the author's search for "black girls," the book elucidates the ways search engines incorporate racist and sexist logics into information systems. Her work shows how the design of information systems, particularly search algorithms, do not just store, sort, and distribute data but also reproduce patterns of inequality. At the beginning of her research, when she typed "black girls" into a Google search box, Noble was faced with pornographic imagery and tropes about black women's "sass" and anger. This contrasted with searches for "white girls," which displayed images of innocence and childhood. Far from objective, racist and sexist search results are at once a function of cultural norms and technical design. With algorithms trained on search terms and clicks from socially situated users, the patterns, prejudices, and problems that persist in the culture are encoded into Google's information infrastructure.

Eubanks, Noble, and other critics reveal the politics of design so that we may fix evident problems, create better technologies, and work toward building a better society.[16] As Noble argues, "the more we can make transparent the political dimensions of technology, the more we might be able to intervene."[17] The practical turn in design studies takes up the task of building better, more ethical, and more equitable things.

The practical turn in design studies is premised on the idea that recognizing values and ethics in technologies will expose problematic politics and enable designers to effect change. The practical turn centralizes ethical considerations in technical design decisions. The tradition posits that engineers and technology producers have an opportunity and responsibility to build products and systems that serve the social good—or at least avoid enacting harm. The value-sensitive design research program and Flanagan and Nissenbaum's *Values at Play in Digital Games* are key representative works from the practical turn.

The value-sensitive design research program is dedicated to constructing methods of making by which producers remain sensitive to ethics and values from the first stage of the design process and throughout implementation and distribution. Value-sensitive design centralizes power relations and inequalities in its treatment of technical products and systems. It begins with the understanding that default designs often reflect default status structures. The program thus works to avoid and ameliorate material reifications of inequality.[18]

In *Values at Play in Digital Games*, Flanagan and Nissenbaum take on the project of practical intervention by focusing specifically on games. Their analysis of the way leisure products embody implicit and explicit social agendas highlights the pervasiveness of politics in design. With clear implications for technological design more generally, the authors demonstrate the ways game design can perpetuate or resist intersecting oppressions of race, class, gender, (dis)ability, and social class. They show that technical objects are infused with values such as privacy, autonomy, stewardship, and equality. These values can at times sit in tension with each other and between stakeholders, manifesting in divergent ways for the diverse subjects who play.

Both *value-sensitive design* and *values at play* detail methods by which technology producers can account for value tensions and engage in socially intentional design practices. These methods include concrete strategies such as identifying direct and indirect stakeholders, collaborating with diverse stakeholders during all stages of production, making incremental changes in the testing phase (for example, by removing or adding a single feature at a time), externalizing values through sketches and scenarios, prototyping, and creating coding manuals with value orientations. Thus, the practical turn takes a critical perspective on technology and addresses this perspective in material form.

The mechanisms and conditions framework of affordances effectively serves both political analysis of technologies and design-based intervention. The automated decision systems detailed by Eubanks can be presented as *refusals* against poor citizens to maintain privacy and *demands* on welfare recipients to accept monitoring. Eligibility standards construct rigid depictions of responsible and deserving subjects, and the automation of these decision systems strips away the human element. Thus, although eligibility standards have traditionally *requested* that recipients comport themselves in line with state-determined values, automation strengthens these *requests* into *demands*. These demands of responsible personhood do not apply equally to

everyone but exert greater force over those with deeper entrench-
ment in poverty and state intervention. For instance, automated
child protection algorithms count any interaction with services
as a risk factor for future abuse. Children whose parents are
monitored are entered into the system. When these children
grow up and start their own families, they do so with marks
already against them. State welfare institutions thus *encourage*
all parents to perform (government-sanctioned) responsible par-
enthood, *refuse* to let poor parents deviate, and *demand* compliance and monitoring in circumstances of intergenerational
poverty.

In a similar vein, the information systems described by Noble
in *Algorithms of Oppression encourage* racism under the guise of
objectivity. The systems *demand* curation on the basis of popular-
ity and advertising relevance. Though users are *allowed* to enter
any search terms they wish, the results they receive *discourage*
critical interpretation. Because media literacy and competence
in critical race and gender studies can loosen the constraints of
the Google search apparatus, *dexterity* with Google's search features and an understanding or *perception* of results as subject to
change alter users' relation to the search tool.

Demarcating the conditions under which technical systems
request, demand, encourage, discourage, refuse, and *allow* not only
identifies the politics and values in technical systems but also
lays the groundwork for intentional (re)design. Here the mechanisms and conditions framework operates in service of the
practical turn. Designers and engineers might rework existing
products to encourage gender equity or demand privacy maintenance. They may build goods and services that request socia-
bility or refuse class-based discrimination. The mechanisms and

conditions framework thus emerges as both an analytic tool and as a device for developing desirable outcomes.

In sum, the mechanisms and conditions framework operationalizes "affordance," providing precise language with which to address human-technology relations. This operationalization is both agile and empirically agnostic, meaning it is not tied to any particular technology but is applicable across myriad sociotechnical systems. The framework can equally address the mechanisms and conditions of bots, social media platforms, chalkboards, seat belts, and shopping carts. This flexible orientation gives affordance analyses both breadth and longevity. One of life's few inevitabilities is that things change, and technological change persists with striking rapidity. Keeping up with sociotechnical change means creating analytic tools that move along with subtle and dramatic technological shifts. The mechanisms and conditions framework is thus transferable by design.

Outline of the Book

The book follows a trajectory from history and politics to conceptualization and methods. Each chapter builds on preceding chapters. However, each chapter is also self-contained and most can be read independently. The only exceptions are chapters 4 and 5, which explicate the mechanisms and conditions framework in detail and should be read together.

The book begins with a brief history of affordance as a concept. One sign of a successful concept is its application across fields. Affordance has certainly achieved this feat. The concept of affordance originated in ecological psychology and has since migrated to design studies, science and technology studies (STS), communication studies, education, anthropology, sociology, engineering, and elsewhere. In its migration and application, scholars and practitioners have undertaken extensive theoretical reworking and engaged the concept in myriad empirical studies. Chapter 2 weaves the varied threads of affordance's intellectual history into a legible and coherent story.

Chapter 3 gives theoretical grounding to the political nature of the mechanisms and conditions framework. Tracing back to media studies scholars of the 1950s and coming up through contemporary STS perspectives of the new millennium, chapter 3 distinguishes affordance analyses from actor-network theory (ANT)[19] and situates it instead with the critical approach of technology as materialized action.[20] Central to this critical framing is an asymmetrical relationship between subjects and objects and a distinction between technological efficacy and human agency.

Chapters 4 and 5 lay out the mechanisms and conditions framework. Chapter 4 explains and exemplifies *how* technologies afford through a porous continuum of *request, demand, encourage, discourage, refuse*, and *allow*. Chapter 5 looks at the dynamic relationship between subjects and objects and their contextual contingencies through the conditions of affordance. It demonstrates how the mechanisms of affordance take shape through variations in *perception, dexterity*, and *cultural and institutional legitimacy*.

Chapter 6 takes up methodology. The mechanisms and conditions framework is an analytic tool. Chapter 6 addresses existing methodological approaches that pair well with this analytic tool. The chapter is geared toward putting affordance analyses into action. The chapter is also of theoretical relevance because it clarifies the criteria by which methodological approaches fit within the scope of the mechanisms and conditions framework.

In clarifying these criteria, chapter 6 rehashes key tenets of the mechanisms and conditions framework and its underlying assumptions.

In the conclusion, I suggest some big questions for future research. The conclusion is meant to be a springboard from which the mechanisms and conditions framework can take flight. My goal throughout the book is to theorize affordances in a way that simplifies rather than complicates. In the conclusion, I urge researchers to apply the mechanisms and conditions framework to the arduous tasks of both analysis and design.

2 A Brief History of Affordances

When I began thinking seriously about affordances, I often
stated that the concept was undertheorized. This is a common
declaration from scholars who write about the topic, and it was
appealing to me as a justification for my own work. If affordances
were undertheorized, then perhaps I could make a meaningful
and substantial contribution to the field. The claim also seemed
empirically true. There are reams of academic texts that use the
term *affordance* as a central analytic device but provide no definitions, further explication, or serious attention to its theoretical underpinnings. Yet the more I read, the less comfortable I
became with my own assertion.

As I came to discover, the scholarly treatment of affordances
has been extensive and sophisticated. I found myself buried
under piles of literature, much of which is painstaking and
detailed. Specific relationships between artifacts, subjects, and
environments have been formalized through numeric equations,
careful nomenclature, graphs, charts, arrows, and appendices.
Debates have been robust, and word counts expansive. Affordances are, in short, very theorized. At the same time, however,
there remains definitional confusion, conceptual looseness, and

an oddly accepted convention of using the concept as though it has no intellectual history at all.[1]

Paradoxically, the affordance theoretical literature is dense and unwieldy, and yet in practice, it is apparently ignorable. I wonder whether this contradiction is more than just a fluke. The strength of affordance as a concept is its efficient manner of expressing technological efficacy without falling into determinism. Its beauty is in its parsimony. A theoretical trajectory that overspecifies affordances and related conceptual variables (including artifacts, environments, organisms, users, designers, and architectures) may obscure, rather than reveal, the concept's full potential. Disciplinary jargon doesn't help, either.

After immersing myself in fifty years of affordance literature, I now contend that the concept needs not more theory but smarter theory. Affordance needs a theoretical treatment that does justice by its richness and depth while maintaining the simplicity that makes the concept an elegant and practical tool. This is my aim with the mechanisms and conditions framework, presented in chapters 4 and 5. To get to the framework, the first task is to lay out and untangle affordance's conceptual history. Such a project sets the foundation for my own conceptual model and also highlights the rigorous and thoughtful work that already exists, bringing together multiple threads into a legible and coherent whole. This chapter offers a foray into the main ideas, debates, and applications of the concept since its inception in the 1960s. Rather than a complete catalog of affordance references, I focus on the most influential pieces and those that most clearly demonstrate relevant lines of thought. This is not an exhaustive literature review but a narrative about where affordances have been and how they can be mobilized for both analytic and practical purposes.

Origins in Ecological Psychology

James J. Gibson first introduced affordances as the pinnacle concept in his work on direct perception.[2] An ecological psychologist, Gibson departed from the dominant perspective of the time, which emphasized representation and inference. Rooted in the ideas of nineteenth-century German scientist and philosopher Hermann von Helmholtz, psychologists in the 1960s predominately modeled perception as a three-term system.[3] The three-term model of perception assumes that perception is the function of a subject, an object, and a mediated representation. For instance, a person (subject) sees a tree (object) via a representational image on the retina (mediator). The subject uses existing knowledge to disambiguate the mediated image and make sense of it.

Gibson rejected this representational perspective in favor of a two-term model that includes only objects and subjects (or as Gibson would say, environments and organisms).[4] The representational model was referred to as inferential perception, whereas Gibson was interested in direct perception. Inferential perception requires that representations are disambiguated via subjects' existing knowledge. Gibson argued that subjects do not need existing knowledge of a situation to disambiguate but instead can perceive directly from the environment and act based on direct perception. That is, the predominant view of perception at the time was inferential and representational. In contrast, Gibsonian perception was direct, antirepresentational, and action-based.[5]

The concept of affordance was central to Gibson's thinking. In 1966, Gibson first defined affordances as "what things furnish, for good or ill."[6] A decade later in his now canonical text

The Ecological Approach to Visual Perception, Gibson expanded the definition:

> The *affordances* of an environment are what it *offers* the animal, what it *provides* or *furnishes*, either for good or ill. The verb *to afford* is found in the dictionary, but the noun *affordance* is not. I have made it up. I mean by it something that refers to both the environment and the animal in a way that no existing term does. It implies the complementarity of the animal and the environment.[7]

For Gibson, affordances are action-based, dynamic, and necessarily relational. Perception is a direct dispositional relation between objects and subjects in which opportunities for action are the driving force. For instance, Gibsonian affordances are not concerned with the Euclidian space between points but instead with the distance between points in relation to a subject's stride.[8] "[W]hat we perceive when we look at objects are their affordances, not their qualities," says Gibson.[9]

Gibson's ideas stem from gestalt psychologists who were working in the 1930s, especially Kurt Lewin and Kurt Koffka, who were interested in perception and sensemaking as greater than the sum of individual parts.[10] For instance, Koffka describes mailboxes as having a "demand-character" for those seeking to mail a letter. That is, the mailbox is not just its material elements, but the materialization of an action opportunity for a subject in need. Gibson builds on this by arguing that affordances are action opportunities that derive from a relationship between properties of objects and properties of subjects, regardless of the subject's need or propensity.

Gibson's conceptualization of affordance has two critical elements: objectivity and bidirectional relationality. Affordances are opportunities for action, based on both intrinsic properties of objects and their relation to subjects. That is, affordances are opportunities for action, not necessarily their actualization. As Gibson explains, "an affordance is not bestowed upon an object by the need of an observer and his act of perceiving it. The object offers what it does because of what it is."[11] Of postboxes and letter writing, Gibson says:

> For Koffka, it was the *phenomenal* postbox that invited letter-mailing, not the *physical* postbox. But this duality is pernicious. I prefer to say that the real postbox (the *only* one) affords letter-mailing to a letter-writing human in a community with a postal system. This fact is perceived when the postbox is identified as such, and it is apprehended whether the postbox is in sight or out of sight (emphasis in original).[12]

For Gibson, affordances are not predicated on use but are manifest in relation to socially situated subjects. Objects and subjects are therefore co-constitutive, and affordances are *potential* actions arising from bidirectional object-subject relations.

Gibson's concept of affordance became significant in the psychology of perception. Since then, it has branched fruitfully into a diverse range of fields, where it remains influential to this day. Key expansions have taken hold in design studies and human-computer interaction, anthropology, engineering, communication studies, and education with a focus on pedagogy and technology.

Affordances Spread

The first major shift in affordance theory came in 1988, when Donald A. Norman introduced the idea of affordances to design studies and human-computer interaction (HCI). Norman's eminent work *The Psychology of Everyday Things* (*POET*) contends that objects should be designed in ways that guide users' perceptions and thus guide action.[13] For Norman, an effective designer should also be an insightful psychologist who builds objects in ways that direct users along intentional pathways. He recognizes that objects have multiple affordances and calls on the designer to highlight desired and relevant action opportunities. Norman first defined *affordance* as follows:

> The term *affordance* refers to the perceived and actual properties of the thing, primarily those fundamental properties that determine just how the thing could possibly be used. A chair affords ("is for") support and, therefore, affords sitting. A chair can also be carried.[14]

Norman eventually renamed his germinal work from *The Psychology of Everyday Things* to *The Design of Everyday Things* (*DOET*).[15] Not only does the updated version have a new title, but it also presents new theoretical delineations that attend to critiques against the original text.

In its original formulation, Norman's *POET* emphasizes perception, in contrast to Gibson, who speaks of the inherent properties of an environment. Critics argued that Norman's formulation gives short shrift to materiality. It is too subjective, they said, and does not grant enough efficacy to material conditions.[16] A decade later in *DOET*, Norman addresses this point by distinguishing between "real" affordances and "perceived" affordances. Real affordances are the actions that an environment makes available, and perceived affordances are those that the user knows are available. He argues that this is a key distinction and that designers should focus on the latter.

In the updated text, Norman envisions object-subject interactions as a series of distinct constraints. He differentiates between cultural constraints, physical constraints, and logical constraints. Physical constraints are synonymous with affordances, logical constraints are what the design environment makes readily available, and cultural constraints are norms shared by a group. Referencing cultural constraints, he further differentiates between affordances (real and perceived) and conventions. Conventions are cultural constraints that have evolved over time, encouraging some actions while inhibiting others. He summarizes the updated argument as follows:

> Affordances specify the range of possible activities, but affordances are of little use if they are not visible to the users. Hence, the art of the designer is to ensure that the desired, relevant actions are readily perceivable.[17]

Gibson and Norman both convey an image of objects and subjects in relation. However, their work derives from distinct disciplinary traditions, each maintains unique purposes, and each diverges from the other in the primacy of objectivity (Gibson) and subjectivity (Norman). Norman's distinction between real and perceived affordances works toward reconciling the two formulations, but daylight remains between these foundational statements on the concept. Drawing variously from Gibson and Norman, the concept of affordance has found its way into myriad fields outside of psychology and HCI. Indeed, disciplinary expansion of the concept appears in anthropology, engineering, communication studies, and education, with threads seeping into neuroscience, robotics, sociology, and philosophy.

Although affordance's interdisciplinary spread has resulted in a dense and at times unruly literature, it also demonstrates the potential for the concept as an analytic tool that spans disciplinary boundaries. Such tools are critical in a historical moment marked by rapid social and material change. Contemporary problems are increasingly beyond the scope of singular disciplinary expertise. Yet true interdisciplinary collaboration is often stifled by distinct languages and conventions that create barriers to communication and understanding. A concept that has organically traveled from one discipline to the next demonstrates strong potential as an intellectually unifying force.

Anthropologists have adopted affordance as a means of cross-cultural understanding and analysis.[18] By rejecting the assumption that humans are distinct in their reliance on symbols and accepting instead the premise of direct perception, anthropologists can learn about new cultures through shared perception (the affordances of shared place and space) and can analyze cultures outside of their own without the troubling distinction between "us" and "them." Tim Ingold explains:

> The argument, in a nutshell, was that a relational approach to affordances might give us a language in which to express how people continually bring forth environments, and environments people, that could escape the endlessly self-replicating dualism between a universally given world of nature and the diversely constructed worlds of culture.[19]

In this vein, Bryan Pfaffenberger advocates for affordance as a conceptual means to capture the tridimensional relationships between technique, sociotechnical systems, and material culture.[20] Through affordances, anthropologists have a dynamic way to understand the interplay between the resources with which artifacts are made (skills, knowledge, and tools), the sociotechnical systems that link cultural practices with technological developments, and the tangible material culture that results from and cycles back to inform cultural praxis. Thus, an anthropological observation of public transit behavior in Beijing would account for the interplay of urban infrastructure, population density, and cultural sensibilities as cocreating both objects (trains, platforms, buses, and share bikes) and subjects (commuters, tourists, and private motorists). The affordance perspective gives the anthropologist an analytic lens with which to understand people and culture in context.

In engineering, Jonathan R. A. Maier and George M. Fadel have led the field in constructing an affordance ontology and method of implementation.[21] Their *affordance-based design* (ABD) introduces affordances as fundamental to engineering design and defines *affordance* as the relationship between two subsystems in which potential behaviors can occur that would not be possible with either subsystem in isolation. ABD incorporates four basic elements: artifacts, users, environments, and designers. Affordances are the relationship between artifacts, users, and environments. The job of the designer is to optimize the intersection of these three elements toward some defined goal or goals. This resonates with Norman's original call for adequate "mapping," in which designers are tasked with psychological insight as they build technologies that clearly guide users down intended paths. As a simple example, chest-height desks facilitate standing, and waist-height desks are primarily suited for sitting. The former guides users down a "healthy" and active physical relationship to the workspace, whereas the latter guides users toward stagnation. An active stance is thus likely preferable if the goal is health and wellness. A sedentary disposition may be preferable if the goal is long stretches of uninterrupted productivity. With an affordance frame, engineers can design with these (and other) various goals in mind.

Within communication studies, affordance has emerged as a robust concept in the study of information communication technologies (ICTs) and computer-mediated communication (CMC). Affordance is useful for its capacity to capture the ways hardware and software interact with socially situated users.[22] Affordance research in communication studies shows how digital architectures, infrastructures, policies, and practices shape and reflect social dynamics. Hence, a review of affordances in the ICT/CMC literature shows studies variously emphasizing design architectures,[23] individual user practices,[24] platform policies,[25] and informal conventions.[26]

Digital and electronic media have also driven the conceptual use of affordance in studies of education and pedagogy.[27] Scholars contend that educational technologies interact with learners to construct learning environments with greater or less pedagogical value. For instance, Roy D. Pea utilizes affordance to conceptually describe the interplay between students and technical systems in a distributed learning environment.[28] Diana Laurillard and colleagues contend that affordances can shape the relative learning benefits for experts and novices in diverse learning groups.[29] Daniel D. Suthers explores how learning goals can be designed into technical systems,[30] and Grainne Conole and Martin Dyke tease out the criteria for technological affordances that enable collaborative learning.[31]

In short, affordance has conceptual legs, and those legs have traveled. The concept now spans multiple fields and does diverse and important analytic and practical work. The immense breadth of a single concept speaks to its hardiness. And yet the concept has not been without controversy. Indeed, the affordance literature is thick with debate and critique, much of which revolves around various emphases on objects, subjects, and contexts.

Objects, Subjects, and Contexts

Since Gibson introduced affordances in the 1960s, the literature has been active with debates about the primacy of subjects versus objects and about the role of context and culture in affordance analyses. Gibson's antirepresentational direct perception approach positioned affordances as bidirectional relationships between "organisms" and "environments." However, some interpret his definition ("what things furnish, for good or ill") as a model in which environments have disproportionate weight while organisms respond only to environmental stimuli. In contrast, critics point to Norman's conceptualization as overly perceptual, unable to adequately attend to material features outside of what subjects perceive. Debates within the affordances literature thus posit various ways to portray object-subject dynamics most precisely. Moreover, analysts contend that neither Gibson nor Norman fully account for contextual and cultural factors. Critics thus build on early works by advancing models of affordance that situate objects and subjects within sociostructural conditions.

Although Gibson's conceptualization of affordance is ontologically bidirectional, defined as a relation between environments and organisms, his work is largely concerned with how the environment emerges as directly perceivable. Thus, his work has been interpreted as maintaining an emphatic bias toward objects rather than subjects.[32] Seeking to rectify Gibson's materialist leanings, the psychologist William H. Warren recentralized subjects in affordance analysis through a case study of stair climbing.[33] Warren set out to determine the relational properties that make stairs unclimbable, climbable, and optimally climbable for distinct subjects, so he quantified the relationship between leg length and riser height as a metric for stair climbability. Not only did Warren show how the properties of objects (riser height) and properties of subjects (leg length) exist in relation, but he also demonstrated subjects' active perception when interpreting the objects with which they engage. Warren's subjects showed remarkably accurate perception of the ease or difficulty with which they would be able to climb a set of stairs, indicating the relevance of perception in object-subject relations. Warren's case study remains a quintessential example of affordance relationality that contemporary theorists continue to evoke.

A group of philosophers built on similar ideas to those advanced in Warren's stair study and introduced *effectivity* as a conceptual way to balance out Gibson's theorizing.[34] Effectivity was set up as a complementary concept that emphasizes subjectivity in perception and the capacity to act. Thus, "The animal's effectivities are directed to the environment in the way that the environment's affordances are directed to the animal."[35] The effectivity-affordance duality ensures equal and dynamic relations between subjects and objects.

Although effectivity deals with the issue of relationality, critics contend that it undermines the power of Gibson's concept, which explicitly entwines environment and subject. Constructing two complementary concepts (affordance and effectivity) thus undermines affordance's bidirectional quality, which is its most crucial feature.[36] Nonetheless, the effectivity-affordance duality maintains purchase within ecological psychology and was formalized by Michael Turvey with a focus on actualization. Turvey contends that affordances are not ontologically present in the environment and that effectivities are not ontologically present in the subject. Rather, affordances are actualized through the *match between* particular object affordances and subject effectivities.[37]

Another conceptual distinction that has emerged is between *utility* and *usability*.[38] This is an effort to capture the materiality of Gibsonian affordances while addressing the perceptual focus of Norman's work. The utility of an object refers to its potentialities in relation to subjects, while usability refers to the perceptual information signaled to the subject by the object. A similar distinction has been introduced in engineering through the complementary relationship between *functions* and *affordances*.[39] Functions are those features designed into an object, while affordances are the "totality of behaviors the user can perform with it."[40] Again, we see a relationship between material potentialities and subjective perceptions that affect—but do not determine—actions and outcomes. These ideas are further expanded as theorists take on the additional variable of context.

In addition to efforts toward reconciling objects and subjects in affordance analyses, theorists have also endeavored to account for context. Anthony Chemero contends that in order for an affordance to actualize, there must be a fit between the properties of the object and the properties of the subject, *along with circumstances that support perception and enactment.*[41] In this way, a meshing of object and subject does not determine an outcome but generates a potentiality that can change across time, between subjects, and amid new circumstances. From this perspective, the "affordances of technological objects are not reducible to their material constitution but are inextricably bound up with specific historically situated modes of engagement and ways of life."[42] Building on this, Andrea Scarantino distinguishes between *surefire* affordances and *probabilistic* affordances. Surefire affordances manifest in a certain outcome, and probabilistic affordances have a positive probability of less than 1.[43] That is, under certain conditions, we can expect objects to elicit a predictable and certain response (surefire), and in other conditions, the environment will push in one direction, but outcomes are not inevitable (probabilistic).

Tied up with contextual factors is the social element of technological artifacts. Neither objects nor subjects exist in isolation. Rather, objects and subjects are part of a world that is "propertied by other people"[44] and by other things.[45] Capturing the social element, the term *social affordances* theorizes an intersubjective relation between persons in situations that shape the meanings, perceptions, and affordances of physical objects.[46] Richard C. Schmidt demonstrates social affordances using the example of a cup with a handle. The cup takes on one meaning as an object for purchase in a store and yet another when given as a gift. It thus affords grasping, filling, and drinking-out-of but also affords capitalist exchange, relationship building, and memory making.[47] In this vein, *organizational affordances* capture the ways organizational bodies interplay with technical systems to shape one-to-many and many-to-many interactions and relational dynamics.[48]

Summarily, Gibson originally conceived affordances as something that "cuts across" object-subject relations. Norman then applied the concept to HCI, merging the roles of designer and psychologist. The concept was and remains influential. However, analysts found early formulations unsatisfactory in their overemphasis on either materiality or perception. Attempts to rectify the issue generated complementary concepts such as *effectivity*, *function*, and *utility and usability*, all of which capture the relevance of perception and its imbrication with materiality as affordances take shape and animate action. The role of context has also risen to the fore with contentions that objects and subjects are enabled and constrained through cultural conventions, social relationships, and situational factors that shape meaning and action opportunities.

Sustained Critiques

Affordance has enjoyed conceptual longevity and proven analytically useful across multiple disciplines. Despite or perhaps because of this, the concept has also endured sustained patterns of critique. Three main critiques are leveraged against the concept of affordance: definitional confusion, binary application, and failure to account for diverse subjects and contexts. As demonstrated in the section above, analysts have certainly worked to address each of these issues. However, the critiques have yet to be resolved in a systematic or widely applicable way.

If you speak with people who study affordances, there is a high probability that they will lament the concept's misuse, overuse, and entirely undefined use within academic literatures. The problem of definitional confusion in affordance analyses is polemic. On the one hand, the concept has been reformulated to death and tied to increasingly specific disciplinary jargon. On the other hand, the concept is often used without any definition at all, as though it has no intellectual roots or any controversy about its meaning.[49]

The seeds of definitional discord may have been sown into Gibson's original conceptualization, in which he advanced "two, apparently irreconcilable positions,"[50] asserting that affordances are intrinsic to the physical properties of an object and at the same time exist only in relation to a subject. Affordances were thus originally conceived as both objective and relational. Movement of the concept from ecological psychology and its reformulation at the hands of Donald Norman exacerbated conceptual uncertainty. Indeed, reviews of the literature on affordance show divergence between definitions derived from Gibson, definitions derived from Norman, and most troubling, use of the term as a central analytic device with no definition at all. Such definitional confusion has become so problematic and widespread that Norman himself has suggested replacing the concept altogether and using "signifier" instead.[51]

A second critique of affordance is its binary formulation in which objects either do afford or do not afford. Despite early works that emphasize the operation of affordances in "degrees," such as Warren's well-known and often cited stair example, practical applications of affordance analyses often depict affordances as either entirely present or entirely absent.[52] Binary depictions not only undermine the concept's analytic integrity but also weaken its capacity as a design tool. Indeed, to capture and evaluate the nuanced interplay between designed objects and user-subjects requires vocabulary that describes affordances that exist between optimal and critical points.[53]

A third critique of affordance is the continued struggle to account for diverse subjects and contexts. Affordance analyses too often describe artifacts as though they exist in a static and monolithic world. This is a somewhat ironic problem, in that affordance was originally formulated to capture a dynamic object-subject relation. That objects afford *in relation to a subject* integrates a notion of variability across persons and contexts. Yet in practice, analysts evaluate objects as though their features are inert.[54] Such rigid analyses deflate a key strength of the affordance concept by undoing its capacity to capture dynamism between subjects and objects within complex and changing circumstances.

Pathways Forward

From its origins in ecological psychology, affordance has spanned disciplines and animated robust debate and critique. It was first formulated as an antirepresentational theory of direct perception that contested dominant assumptions about the relationship between organisms and environments. As it moved to design studies, the concept tasked the designer with the responsibilities of the psychologist and placed deep emphasis on guiding user perceptions. Subsequent advances have worked to add precision to the concept and find balance between materiality and subjectivity. Even with these theoretical advances, the term remains plagued by critique, with central intellectual figures suggesting that we do away with the concept altogether. Yet affordance maintains a strong presence across literatures and shows no signs of waning. It is thus advisable that we attend to affordances in a thoughtful manner rather than tossing up our hands and letting the concept take on a life of its own.

Affordance has been subject to critique over conceptual clarity, binary formulation, and static depictions of persons and contexts. Although each of these issues has received significant attention, there is yet to be a systematic framework that addresses them together in a readily applicable way. A key reason for this is that theories of affordance have remained conceptually siloed within specific fields and articulated through discipline-specific jargon. Even as theoretical advances continue, these advances often remain inaccessible outside of niche academic circles. What is needed is a simple and systematic framework of affordance, articulated with vocabulary that cuts across disciplinary boundaries. Building such a framework begins by taking note of the most useful developments within the affordance literature.

Of the three main critiques, conceptualization has been the most effectively addressed and theorists have done so in ways that correct for binary and static applications. Conceptual advances formulate affordances as continuous (rather than binary) and dynamic (rather than static). The work of Peter Nagy and Gina Neff[55] and Sandra K. Evans and colleagues[56] stand out in this regard. Rooted in communication studies, the conceptual clarifications offered in these works can be applied across fields. Nagy and Neff make the notable contribution of accounting for "webs of relations" between artifacts, users, designers, and contexts in their introduction of *imagined affordance*. *Imagined affordance* is an interplay of materiality, intentionality, and serendipity as designers build objects that then take shape through diverse users and changing circumstances. Similar work has emerged in engineering, with scholars articulating affordance relationships between artifacts and each other as artifact-artifact affordances (AAAs), between artifacts and users as artifact-user affordances (AUAs), and artifacts in environments as artifact-environment affordances (AEAs).[57] Adding further precision, Evans and colleagues articulate an affordance as that which mediates between features and outcomes. This formulation attends to materiality (features) while recognizing the myriad ways in which materiality can manifest through socially situated subjects, resulting in a range of undetermined outcomes. Thus, affordances are potentialities that operate in degrees through interactions with diverse subjects and circumstances.

Building on these recent advances, the mechanisms and conditions framework provides a common language, untied from disciplinary jargon, that recognizes affordances as both gradated and contextually situated. First introduced as a simple tool that cuts across disciplines to enable dynamic sociotechnical analyses,[58] the framework is already being put to use across diverse fields. Scholars have employed the framework to understand complex object-subject relations, account for diverse user practices, and address structural power relations.[59] It has even extended out from technology studies to serve as a framework for broader patterns of power-infused interactions.[60] I further articulate the mechanisms and conditions framework in the remaining pages of this book, constructing a foundation for affordance analyses moving forward.

3 Politics and Power

The social world is power laden, and so too are technologies. Conceptually tools for the study of human-technology relations must therefore also assume and attend to political dynamics as they manifest in social and material forms. This chapter follows major lines of thought in the evolution of communication and technology studies and situates the mechanisms and conditions framework within and against them, highlighting the model's critical orientation. The chapter establishes two key assumptions: humans and technologies are co-constitutive, and politics and power are central to this sociotechnical relation.

Conceptually, affordances address the shaping effects of technologies in a way that avoids technological determinism. Technologies may affect human life in myriad and sometimes profound ways, yet outcomes are never certain and can be disrupted, thwarted, and circumvented to sometimes surprising ends. That is, both humans and technologies are powerful, protean, and eventful.[1] This perspective resists designations of *either* human subjects *or* technological objects as autonomous and effectual and instead positions human-technology dynamics as necessarily relational.[2] The mechanisms and conditions framework thus assumes that technologies and people exist together in co-constitutive assemblages.

Most science and technology studies (STS) scholars today assume co-constitutive assemblages as a starting point for analysis. Human-technology relations are intrinsically relational. My arguments diverge from predominant perspectives, however, by establishing agentic asymmetry between human subjects and technological objects. I contend that although humans and technologies mutually construct each other, the weight of responsibility always falls to people. This does not mean that humans have disproportionate effect. Indeed, technologies may shape the world in ways humans could never dream and at magnitudes far beyond the capacities of mere flesh. Rather, the assumption of asymmetry is based on distributions of accountability. Technological objects can exert substantial force, but only humans can and must be held to account. I hinge the assumption of object-subject asymmetry on a distinction between *efficacy* and *agency*. Efficacy refers to the capacity to effect change. Agency refers to the capacity to inflict will. This distinction comes from Ernst Schraube's *technology as materialized action*[3] approach, which claims that although technology can be highly efficacious, only humans can be agentic.

I build my argument by drawing on three key lines of thought: Marshall McLuhan's classic thesis on the medium as the message,[4] Bruno Latour's actor-network theory (ANT),[5] and Schraube's notion of technology as materialized action.[6] I also pull from Langdon Winner's delineation of artifacts and their politics as foundational evidence of how power relations permeate sociotechnical systems.[7] I conclude by proposing the mechanisms and conditions framework of affordances as a neat analytic tool that captures technological efficacy and holds it together with human agency, always accounting for contextual variation and looming structural hierarchies. In short, this chapter describes how technology is efficacious, political, and inextricable from the human element.

The Medium Is the Message: McLuhan on Technologies as Objects of Study

Scholars within communication studies and STS have taken important strides to demonstrate the shaping effects of technology. Analysts make a compelling case that technologies do things, and as researchers, we should take those things seriously. This is the key contribution of communication scholar Marshall McLuhan, who in 1964 famously declared that the medium *is* the message.[8] McLuhan was pushing back against what he saw as two recurrent and related errors in academic commentary on media in society: (1) the presumption that technology is neutral and (2) an exclusive focus on media content as the unit of analysis.

McLuhan directly opposes the idea of technological neutrality. The assumption that technology is neutral means that the technology itself has no organizing function and instead, all that matters is what people do with technological objects. A position of technological neutrality, or extreme constructivism, ignores a deep empirical history in which social life has continually reformed in the face of technological change. For instance, the industrial revolution brought with it not only changing work conditions but also the rise of cities, the emergence of a "middle class," and a restructuring of families that tied many women to uncompensated labor in the home. In this way, the introduction of the train rail organized economic and political life around periodic stops on a fixed geographic trajectory, and automobiles rearranged the social infrastructure around complex and interweaving road systems. The rail system fostered centralized towns, and automobiles enabled the development of suburbs and freed commerce from the rigid temporal and geographic constraints of rail tracks and train schedules.

As a communication scholar, McLuhan was primarily concerned with communication media like newspapers, telephones, radio, and television. Just as railroads, cars, and industrial machinery are not neutral, neither are the technologies through which we produce and consume information. For McLuhan, the job of the communication scholar is to understand the social underpinnings and implications of communication media, including whose interests they serve and how they might be resisted.

Related to the fallacious assumption of technological neutrality, McLuhan critiques an overemphasis on *content* within media studies. Following World War II, media and communication scholars became preoccupied with powerful broadcasters and their potential influence over individuals and publics through implicit and explicit propaganda. This concern gave rise to the "media effects" paradigm in which media products are studied as forces of cultural construction.[9] McLuhan advocates for a shift away from media content and a shift toward media proper.

McLuhan argues that analysts should look beyond what people produce and consume through a given medium and instead try to understand the medium itself. It is the medium, claims McLuhan, that has significant effects on individuals, cultures, and the rhythms of public life. That is, the medium *does* something in its own right and should thus be the primary object of analysis. In other words, the medium is the message.

Distinguishing between *medium* and *content*, McLuhan explains that the former is a technological apparatus and the latter includes the range of outputs from that apparatus. Using electric light as an example, McLuhan refers to the light itself as the medium and the illumination from varied sources—including reading lamps, surgical lamps, and televisions—as the content. It is crucial to McLuhan that scholars focus on the medium rather than be distracted by content. He argues that content can take myriad forms and is largely irrelevant. The medium is what shapes society and should thus be the object of scrutiny.

Positioning himself against the prevailing perspectives of the time, McLuhan suggests that focusing on content ignores the power with which technology affects individual lives and collective social organization, obscuring the forest for the trees. Understanding television based on programming, food systems based on grocery store shelves, or social media based on the substance of newsfeeds would all be examples of content-focused analyses. Instead, McLuhan would have us interrogate the infrastructure of television streaming services, the technologies of mass food production, and the principles of algorithmic networked sociality. For McLuhan, understanding mediated technologies is not about analyzing what people produce and consume through them but discerning each medium's syntax and grammar

McLuhan warns that myopic attention to outputs—or content—obfuscates the ways media infiltrate the fabric of daily life. Asking what people do with technologies displaces the bigger issue: what technologies do with people. McLuhan argues that once introduced, media quickly become entrenched. People are then swept away in the medium without an opportunity to put on the breaks or change direction. Maintaining social autonomy, then, requires a critical eye toward technological objects and the media systems in which they are embedded.

McLuhan cautions that ignoring media's shaping effects fosters naivety and leaves people vulnerable to mechanisms of control over which they have little recourse. He thus provocatively states: "subliminal and docile acceptance of media impact has made them prisons without walls for their human users."[10] McLuhan's warning seems especially pointed in the face of algorithmic systems increasingly charged with critical functions such as hiring decisions, public resource allocation, criminal justice outcomes, knowledge curation, and information distribution. Recognizing the medium as the message is McLuhan's key to avoiding pervasive technological constriction.

Actor-Network Theory: Overcoming Technological Determinism

McLuhan's contributions are intellectually important as a counter to extreme constructivism and a reorientation toward technological efficacy. In response to the common adage "Guns don't kill people, people kill people," McLuhan would likely respond, "Guns generate systems of violence." McLuhan's insistence on the medium as the message casts light on the ways that technical infrastructures shape social life. However, his work has been critiqued for its technological determinism. Technological determinism means that technologies prefigure (or determine) a range of effects. Critics point out that people are not simply dupes upon which technologies act, but are active subjects who creatively engage in technological implementation and use. Although McLuhan made a significant contribution by reminding people that the medium matters, critics argue that he takes the case too far and erases human agency.

Actor-network theory (ANT) arose in response to the technological determinism promulgated by McLuhan and his contemporaries. Most famously articulated by Bruno Latour, ANT depicts humans and technologies as mutually shaping entities that together make up multifaceted webs of relations.[11] ANT takes seriously the idea that technology is powerful but understands humans as equally so. Just as technologies shape people and societies, people and societies actively build and use technologies. For example, Google Maps does not unidirectionally determine geographies but reflects existing ways of knowing and navigating space and place at the same time that it adapts to users through the collection and deployment of geolocational metadata.

Key to ANT is the idea that humans and technologies engage in mutually constitutive networks or "assemblages,"[12] with no preference or distinction between people and things. All members of the network are considered actants, and actants all combine to create an assemblage. ANT uses the term *actants* to overcome the divide between humans and nonhumans within relational assemblages. *Actant* replaces the term *actor* because *actor* generally has a human connotation. For ANT theorists, human and nonhuman actants are always part of a mutually constitutive actor network. This means the actor network that makes up a classroom setting includes students, teachers, desks, dust, computers, lecterns, and temperature control units. The presence and behavior of all actants make up the classroom experience. Changes, additions, or removals alter the classroom experience. For example, the students, teachers, and computers may become disturbed if the dust participates with too much gusto, and the desks, computers, and lecterns remain restful if the students and teachers decide not to attend class.

Applying the language and logic of ANT to a 2011 Occupy Wall Street protest, technology analyst and STS scholar David Banks describes the process of acquiring wifi for an event in Albany, New York:

> After several hours, the IT working group resolves that 4G hotspots will not cooperate with their encampment. The 4G signal refuses to visit the park with the same regularity as the activists. Without the 4G signal, those in the park are unable to reach their fellow activists, computers, protest signs, and supplies located throughout the Hudson Valley region. The IT working group decides instead, to project a wireless signal from a nearby apartment into the park. They devise an assemblage of signal repeaters and routers that will provide a more reliable stream of data that will show up on time to general assemblies, and in sufficient numbers. The working group believes that the attendance of broadband Internet will allow the geographically and temporally dispersed occupiers to be enrolled within the larger actor-network of Occupy Albany. This increased attendance by activists, broadband connections, and networking hardware, according to the facilitation working group, will lend more authority to the decisions that come out of the GA and keep the occupation going through the winter.[13]

Note how Banks includes human and nonhuman actants as equivalent nodes within the network. The protest is attended by people, signs, and computers. One might say that the protest suffers because 4G is not fully present, just like the protest would also suffer if human activists were unreliable in their commitment to the event or the cause. Luckily for the protesters, broadband and routers actively partook.

ANT is an attractive framework for its capacity to address the meaningful co-constitution of humans and nonhumans. The introduction of *actant* as a piece of terminology and the practice of placing people and things on equal ground effectively communicates that technologies impose on, but do not determine, social and behavioral outcomes. ANT thus captures technologies' shaping effects without getting trapped by technological determinism. For ANT theorists, people and things are part of an integrated and inextricable whole.

The Politics of Artifacts

ANT represents a major advancement in communication studies and STS. It has been and remains highly influential among those who seek to understand and explain the integration of technologies across varied arenas of social life. However, a lingering critique about ANT's struggle to deal with issues of power, politics, and inequality remains a resounding blight on the framework.[14] In this regard, the main trouble with ANT is its symmetrical treatment of all "actants" within a web of relations. All people and things ostensibly play active roles, with no clear guide from ANT to discern which actants hold greater influence, to what ends, and in whose interests.

For critical social scientists, power and inequality are central to the organization of social life. Intersections of race, class, gender, sexuality, (dis)ability, and geography profoundly affect how people move through the world, how they interact with each other, and what opportunities are (and are not) available to them. Through this lens, any social theory that inadequately attends to power dynamics suffers from a serious explanatory deficiency.

Feminist STS scholars argue that ANT's incapacity to address race, class, gender, and other social hierarchies renders the perspective ineffective as a framework for understanding or explaining technology in society.[15] ANT's apolitical foundation precludes the framework from accounting for systems of marginalization and oppression around which social life takes shape. For instance, an apolitical and power-neutral analytic framework would prove wanting when analyzing data-based policing systems that preemptively label individuals and communities as suspect,[16] when examining the development of cinematic technology that optimally captures (and assumes) white skin,[17] or when looking at the data flows in which personal and detailed information spreads from social media platforms to advertisers and political operatives with ethically ambiguous agendas.[18]

Critical scholars contend that power and inequality are endemic to the social system. Thus, any meaningful intellectual approach must address power dynamics. However, contemporary proponents of ANT have marked political analyses as beyond the framework's scope. In a 2014 article clarifying the purpose and tenets of ANT, Edwin Sayes explains that "morality and politics" are outside ANT's parameters. ANT was never meant to account for power, Sayes concedes, and thus the theory should not be critiqued on those grounds.[19] However, critics would say that power and politics are part and parcel of existing social systems. They would say that frameworks with parameters that exclude politics and power are inherently flawed. I concur with this critical take.

The significance of integrating power into frameworks and theories of human-technology interaction is quickly apparent through the now classic work of STS scholar Langdon Winner, who asked the question "Do Artifacts Have Politics?"[20] Winner analyzed the urban planning of New York City with a particular focus on bridges along the Long Island Parkway. Designed by

Robert Moses, the bridges were too low for buses to pass underneath. These low-hanging bridges made the attractive shores of Long Island inaccessible to those who relied on public transit and kept the roads open to people who traveled by car. Public transit disproportionately serves people of low socioeconomic status, which intersects with race such that riders are more likely to be people of color. This seemingly apolitical architectural decision (bridge height) thus perpetuated race-class dynamics in a way that maintained a white affluent demographic on the Long Island beaches, a pattern that remains in place to this day.

Moses's low-hanging bridges are an example of what Selena Savić and Gordan Savičić refer to as the "unpleasant design of 'hostile architectures.'"[21] Unpleasant design regulates social behavior through architectural features that enact control in the absence of authority figures. For example, armrests on public benches deter people from lying down, thus making the space uninviting for homeless populations. In Seattle, the transport authorities have erected bike racks under bridges to displace tent encampments and their occupants. In the United Kingdom, a housing estate mounted unflattering pink lights that show skin blemishes, discouraging teenagers from loitering. Such "hostile architectures" can also take shape through digital design and algorithmic code. For example, automated human resource management programs disqualify applicants without predetermined credentials (or the proper key words), thus disadvantaging candidates with less social capital,[22] and banking interfaces select indicators of who will (and will not) be likely to pay back a loan, thus reinforcing wealth distribution via purchasing potential.[23] In short, technologies are encoded with power relations that produce patterned effects.

Technology as Materialized Action: Technological Efficacy and Human Agency

The main premise of Ernst Schraube's notion of technology as materialized action is that technological objects are imbued with the politics and values of the culture within which they arise. Technologies do not merely mediate between subjects and the world but are material manifestations of subjectivity. Objects maintain a sometimes profound shaping effect, but ultimate responsibility rests with human subjects. For Schraube, "concrete historical experiences, needs, ideas [and] interests . . . flow into the construction of products." In a sense, Schraube's approach adjusts ANT and infuses it with a much-needed critical element.[24]

A central component of the materialized action approach is an asymmetrical relationship between people and things: people maintain a distinct responsibility for the production and use of technological objects. Schraube is clear in his assertion that subjects and objects mutually shape one another. Channeling McLuhan and Latour, Schraube states: "It is not only the subjects that do something with the things; the things also do something with the subjects."[25] However, what distinguishes subjects from things is agency, which Schraube ties to humans exclusively. He explains: "it would be misleading to speak of an object really 'acting.' Action is an intentional human activity accessible to consciousness and concerned with issues of freedom, reasons, and responsibility."[26] Hence, there is a "need for an asymmetrical-reciprocal language" that designates the human as the "responsible acting subject with the potential to engage on a socio-political level."[27] It is from this line of thought that the mechanisms and conditions framework derives its assumption of human-technology asymmetry.

A materialized action approach recognizes technological efficacy (technologies *do* things) but rejects the idea of technological agency. Agency is reserved for human subjects. This distinction between agency and efficacy and the related asymmetry in human-technology relations open the door to critical analyses. Placing agency exclusively with human actors positions producers and consumers as responsible parties. The effects of technology, both good and bad, can be traced back to cultural norms, corporate directives, state interests, and other claims makers and stakeholders. Designers engrain their own agency into technologies, and users agentically employ those technologies. The force of technological objects can be immense, but that force is inextricable from the values, desires, and interests of human actors.

This subject-object asymmetry undergirds the logic behind scholars' treatment of AI as neither artificial[28] nor intelligent,[29] but the material manifestation of human values and biases. Speaking in a similar vein about credit-sorting algorithms, legal scholar Frank Pasquale exemplifies the human origins of seemingly autonomous technological systems:

> Regulators want to avoid the irrational or subconscious biases of human decision-makers, but of course human decision-makers devised the algorithms, inflected the data, and influenced its analysis. No "code layer" can create a "plug and play" level playing field. Policy, human judgment, and law will always be needed. Algorithms will never offer an escape from society.[30]

The practical turn in design studies—discussed throughout the first two chapters of this book—is premised on the idea that human values manifest in technological objects. Human primacy is thus not only a tool of accountability but also an opportunity to make, distribute, use, and refine technologies with intentionally defined value systems. Hence, Peter-Paul Verbeek refers to design as an intrinsically moral endeavor, harking back to Donald A. Norman's original mandate for designers to act as psychologists, guiding users down particular paths and away from others.[31]

To be clear, a theory of technology as materialized action does not presume hand-wringing capitalists who quietly impose their will onto technological objects that then infiltrate the social system through meticulous plots. On the contrary, the effects of any technology remain uncertain, taking shape only through interactions with complex societal structures and diverse users who can deploy the technology toward various ends in sometimes highly creative (and unexpected) ways. Thus, Schraube talks about technology as ontologically ambivalent. He states that "things are more than just societal meanings, more than just socially conceived and produced items. They always materialize, in addition, an unknown action, something coincidental, unplanned, and their decisive power and efficacy can frequently be located just in what had not originally been imagined or intended."[32]

The effects of technological objects may surprise those who make and distribute them. Surprises can derive from creative practices on the part of users, as well as from latent effects that designers and distributors did not foresee or intend. In this vein, the effects of technologies are nearly always multiple, or "multistable."[33] An artifact does not just do something, it does numerous things, many of which were never imagined.

For example, social movements scholar Zeynep Tufekci draws a careful sociological analysis of the role played by digital and mobile technologies in protest movements.[34] She demonstrates that the same technological advancements that enable mass connection and facilitate rapid organization also leave protest groups relatively fragile. Traditional social movements required immense groundwork to establish a presence and organize action. A happy side effect of traditional organizing efforts is that the mundane and tedious processes produce crucial benefits such as group cohesion and clearly defined leaders within the movement. In contrast, digital social technologies help movements grow quickly but struggle to cultivate an infrastructure that can sustain challenges from the state and internal disagreements, rendering movements less solid. The effects of digital tools on protests, then, are multiple and sometimes contradictory. Similarly, the fact that social media are integral to protests and political participation significantly extends the original purpose of some of the most prominent social media platforms. For instance, Facebook started as a social hub meant to connect friends and communities at an elite educational institution. It has now become a key site through which users post abuses by state authorities and document social injustices. It is unlikely that Facebook founder Mark Zuckerberg imagined his platform would host livestreamed videos of US police officers shooting unarmed citizens when he created TheFacebook.com in 2004 or that his team envisaged those shootings when it introduced Facebook Live in 2015.

Technology as materialized action is not so much a negation of ANT as it is an evolution in STS thought. The materialized action approach takes from ANT the clear recognition that technologies and human subjects interact in a mutually shaping relationship. For Schraube, however, the human-technology relationship is asymmetrical. The assumption of asymmetry that underlies the materialized action approach creates space for analyses of politics, power, and human agency. The mechanisms and conditions framework aligns with the materialized action approach, equipping the framework with a critical analytic lens.

Chapter Summary

This chapter establishes two key assumptions that undergird the mechanisms and conditions framework: humans and technologies are co-constitutive, and human-technology relations are power-laden and political. Technologies are imbued with human subjectivity and deployed by creative subjects. The effects of technology can be planned but are never entirely knowable. People may use technologies in innovative and creative ways, and the larger implications of technological developments, however they are used, can be surprising and unexpected. For these reasons, *affordance* is the appropriate terminology for talking about technological objects and their place in sociotechnical systems. The features of the object can be identified, but the uses and outcomes are variable. Objects thus afford but do not determine.

Building on canonical works from communication studies and STS, a materialized action approach fits symbiotically with the mechanisms and conditions analytic framework. This framework of affordances navigates the interplay of technological efficacy along with human agency. In turn, by adopting the human-technology asymmetry engendered in a materialized action approach, affordance analyses hone in on power, politics, and inequality.

The following two chapters offer theoretical precision to affordance theory by operationalizing affordances through the mechanisms and conditions framework. The mechanisms and conditions framework addresses key critiques leveraged against the concept and, in doing so, shifts affordance from a tool that describes *what a technology is* to one that describes *how a technology operates*. This entails the introduction of a clear conceptual model that remains flexible across time, users, and situations, always accounting for structural dynamics.

4 Mechanisms of Affordance

On a chilly day in the winter of 2015, I stood in a classroom talking with the students enrolled in my Cultural Studies of New Media course. The topic of the day was affordances. After a background lecture about the evolution of structure and agency debates in science and technology studies, I introduced the main concept and jumped into examples. My primary objective was to grapple with analytic tensions between technological constructivism and technological determinism. The first example was a fence. A fence does not impose impenetrable borders, I said, but it *affords* spatial restriction. After moving on with a few more examples and some back and forth between myself and the students, a bright young man raised his hand and pointed out that there is a substantial difference between a fence made of wood and an electric fence and that both are distinct from rope fencing. We all agreed and discerned that while the rope fence *asks* you to respect a boundary, the wood fence *tells* you to do so, and the electric fence *insists*.

This student had tapped into a longstanding critique of affordance theory. Although both Gibson and Norman constructed affordances as nuanced gradations, the concept has been applied in a largely binary fashion. That is, analysts who employ the concept do so as though objects either afford or do not afford some function. But like the fence example, objects afford in varying degrees, and their effects are exerted with differing levels of force. The rope asks, the wood tells, and the electricity insists.

Over the years, scholars have tried to get outside of affordance binaries. For example, as discussed in chapter 2, William H. Warren introduced a mathematical formulation to calculate the "climbability" of stairs.[1] The ratio of leg length to stair height makes a set of steps range from optimally climbable to entirely unclimbable, with a series of accessibility variants in between. His theoretical point was that affordances are not present *or* absent but present *and* absent, by degree. Objects do not just afford or not afford but push and pull with more and less pressure. Sandra K. Evans and colleagues highlight this point in their treatment of affordances as mediators between features and outcomes,[2] while Rob Withagen and Harjo J. de Poel point out that affordances are not mere opportunities for action but situationally variable invitations.[3]

Despite empirical and theoretical advances, binary renderings of affordance remain widespread.[4] A binary model of affordance translates to either-or renditions of what an object enables and constrains. Either you can document images with a device, or you cannot; either you can avoid surveillance on a platform, or you cannot; either an object is mobile, or it is tethered in place. For anyone who has engaged with any technology, this either-or rendition is likely inconsistent with experience. Documenting images may be more or less difficult, avoiding surveillance may be automatic or require savvy, and an object may be easy to move, cumbersome to move, or firmly fixed in one location.

I contend that affordance's binary problem is rooted in an entrenched but misguided orienting question. Analysts ask

"What does this object afford?" when the more appropriate question is "How does this object afford?" Altering the question in this small way—from *what* to *how*—reconfigures affordances as continuous and dynamic rather than static and binary. The remainder of this chapter is dedicated to operationalizing the *how* of affordances.

Proper operationalization is critical for transforming a continuous conceptualization of affordance into a practical analytic tool. Expanding on previous work, I suggest a framework in which technological objects do not just *afford* or *not afford* but *request, demand, encourage, discourage, refuse,* and *allow*. Requests and demands are bids placed by technological objects, on user-subjects. Encourage, discourage, and refuse are the ways technologies respond to bids user-subjects place upon them. Allow pertains equally to bids from technological objects and the object's response to user-subjects. Together, these make up the *mechanisms of affordance*.

Before diving into each mechanism, I need to say a bit about how this part of the framework operates. First, the mechanisms are not prescriptive. That is, request, demand, encourage, discourage, refuse, and allow are not concrete categories into which technological features essentially or inherently fall. Rather, these are analytic stopping points that help describe the intensity with which technological objects facilitate or impede particular lines of action and social dynamics. These categories could go by other names, and there could be more or fewer categories. These are meant as a set of hooks on which analysts can hang their descriptions, comparisons, and points of debate.

Related to the flexible (rather than rigid) nature of affordance mechanisms are the porous boundaries between categories. Features may not fit cleanly into one mechanism category or another. Rather, the affordances of an object can slip between categories or rest within the margins. A strong discouragement may also be read as refusal, just as a weak demand may be read as a request. Concretely, this means that analysts and practitioners could reasonably disagree about whether something is a request or a demand, engage in lively disputes about whether something is encouraged versus allowed, and go back and forth internally about whether some action is refused or merely discouraged. This uncertainty is a strength of the framework. It creates a nimble analytic tool that serves—rather than stifles—dynamic readings and renderings of technological objects in society. It also creates a common vocabulary for knowledge sharing, theory building, critique, and debate.

Requests and Demands

Requests and *demands* refer to bids that originate with the artifact. They are initiated by the technological object and guide the user in some direction, with varying degrees of resolve. *Requests* indicate preference for some line of action over others, and *demands* render one line of action inevitable and other lines of action implausible. Although requests and demands originate with the artifact (rather than the user), they are rooted in sociostructural dynamics. Humans design, build, and distribute technological objects and infrastructures. How these objects and infrastructures guide human behavior arises from and is situated within existing social systems.

Requests
When a technology *requests*, it emphasizes a particular set of actions, deemphasizing other action possibilities. A user may

abide by a request, ignore a request, or address it only partially. A request necessarily entails a degree of flexibility. The technology persuades in one direction but leaves alternate options open.

Recalling the fence example from above, we might say that the rope fence *requests* that walkers stay within or outside the perimeter. The rope indicates a preference, but passers-by may easily step over the rope or dip under it. Both the twine around newly seeded grass and the velvet ropes that guide people through long and winding queues shape movement patterns but do so in ways that can be readily overcome. The material of these fences and their arrangement in space can do little to stop someone who wishes to breach the barrier. Thus, the rope fence does not force people out or keep people in but asks them not to intrude or to remain on a designated path.

Continuing with this example, we can see that the strength of a *request* will vary between different kinds of rope fencing, even if the ropes do not differ in physical restraint. For instance, yellow barricade tape adorned with police iconography likely strengthens the force of a rope's request. That is, the police tape makes a stronger request than an unmarked piece of brown twine. Although the materiality of twine and flimsy plastic are not substantially different in their physical capacity to prevent breach, the police tape is bolstered through the semiotics of institutional legitimacy and sometimes actual capacity for punitive measure. The police tape is firmer in its demarcation of a space as off-limits and creates more solid barriers to entry than an equally permeable twine fence without institutional markers.

Variation in the affordances of police tape even persist between jurisdictions due to varying legal ramifications. In 2017, for instance, a Republican member of the Missouri House of Representatives named Galon Higdon proposed (unsuccess-fully) House Bill 37 (HB37), which would make crossing a police border a class A misdemeanor. Breaching a cordoned off area could be punishable by a criminal record, up to a year in jail, and an up to $1,000 fine. Defending the bill, Higdon told report-ers, "Right now, [the police border] is pretty much a request."[5] Apparently, Higdon wanted to move it closer to a *demand*.[6]

The significance of this distinction between plain twine and police tape and between police tape under distinct legal codes is twofold and holds relevance for the mechanisms of affordance more generally. First, it illustrates the fluid and varying nature of affordances. Not all *requests* (or *demands*, *refusals*, and so on) are created equal. The mechanisms are artificial nodes and have room for within-category variation. The police tape veers closer to a *demand*, and the twine rope exerts very little force. Second, the elements that locate a technological object in one category versus another (for example, *request* versus *demand*) are not purely material but take shape in relation to cultural meaning systems and institutional infrastructures. The police tape is no less physically permeable than the twine, yet it enacts spatial restriction more powerfully.

Demands
Requests are distinct from *demands* in the relative availability of alternate options. A request prefers some line of action, but a demand implies there are no other possibilities. Demands exert a strong degree of force. Rather than asking someone to "Please do this, and please do not do that," a demand more firmly states, "You will do this, and you will not do that." A demand might present in the form of physical, social, and/or symbolic prompts.

Returning to fences, ropes represent a request, but steel fitted with electricity represents a demand. An electric fence demands that passers-by remain on one side of the barrier. When navigating space in a prison yard, for example, the fencing structures demand that inmates remain within a clearly defined and bounded space and that members of the public remain outside of that clearly defined and bounded space.

Like fences that organize how people move in space, so too to do roads and rail lines. Highways and train tracks demand that automobiles follow the paths on which the roads and rails were built. We might say that railways generate a stronger demand and roadways lean closer to a request. Not following a train track renders a train dysfunctional, thus making the locomotive technology dependent on the infrastructural technology. Cars remain functional when going "off road," but drivers may suffer vehicular damage, bodily harm, and police sanctions (such as tickets and fines) between points A and B.

In the world of academia, significant attention has been aimed at the distribution (and control) of academic texts. Although digitization creates the opportunity to archive intellectual materials and make them publicly available, many mainstream publishing companies set up infrastructures in which articles are locked behind paywalls. Publishing platforms are then built in a way that *demands* either individual payment or institutional affiliation to access published content. This demand has been the subject of public protest as proponents of open access advocate instead for policies and related digital architectures that do not place financial restrictions on interested publics but instead *allow* knowledge to flow openly and equitably.[7]

On Facebook, the platform continues to *demand* that users select a gender category when signing up for the service. Initially,

Facebook demanded that users select either male or female but has since expanded to include more than fifty custom gender options. That is, Facebook dropped its previous demand that users engage in binary identification but maintains the demand for gender identification of some sort. Facebook also demands that users select from a prefigured list rather than use a write-in box that might broaden the field of self-identification.[8] Platform usage thus requires users to gender identify, but the interface could be (and has been) reconfigured in ways to tighten or loosen those requirements. Facebook's gender-identity demand is a function of its design, and its design is a function of decisions that were neither natural nor inevitable and could certainly be otherwise.

Although *demands* exert force, they are not deterministic. People may opt out of using a technology or may subvert a demand in their use of the technology (though subversion requires significant effort and perhaps a degree of courage and risk). For instance, a person may covertly take a car off road; people may elect not to sign up for Facebook or they may try to confuse the Facebook system by selecting one gender category initially and then signaling alternate gender categories through other fields on the platform; and academics can undercut publisher paywalls through social sites like ResearchGate. Demands thus present as the only possible option but remain vulnerable to unexpected and creative user agencies.

In sum, *requests* prefer, and *demands* insist. Request and demand are not static or uniform categories but represent approximations of the intensity with which a technology pushes users in some directions and pulls them away from others. Within each category, there is room for variation and slippage. A strong request may spill over into a demand, and a weak demand may arguably align with a strong request (that is, other options may seem plausible but unlikely). Together, requests and demands represent the bids technological objects place on users. Although these bids derive from objects, we should be sure to recall that objects are materialized action[9] and thus are inseparable from the sociocultural systems from which they arise and in which they are deployed.

Encourage, Discourage, and Refuse

Encourage, *discourage*, and *refuse* are how technological objects respond when user-subjects initiate some action. These technological responses can accommodate, deter, or block users' initiatives. When technologies *encourage*, they make some line of action readily available and easy to execute. When technologies *discourage*, they erect barriers to a line of action. The action may still be available but not readily so. The user may have to overcome obstacles or creatively engage the technology in order to access lines of action that are *discouraged*. Technological objects *refuse* when some line of action seems entirely untenable.

Like the first set of mechanisms (*request* and *demand*), these three mechanisms are integrated as part of sociotechnical systems involving humans, material apparatuses, culture, and structure. Bids *by* the object (*request* and *demand*) are not empirically distinct from bids *on* the object (*encourage*, *discourage*, and *refuse*). Rather, each serves as a set of analytic pegs that represent distinct foci on particular parts of the human-technology relation. For example, when a technology demands some line of action, it refuses others; when it requests that users engage in some behavior, that behavior is also encouraged.

Encourage

Technological objects *encourage* some line of action when that line of action is made easy and appealing. The action is generally obvious, expected, and seamless to execute. Those lines of action that are encouraged often represent the very things a technology was built to accomplish. Users need to employ little or no creativity, deviance, or subterfuge to engage the technology in encouraged ways. For example, cameras built into phones encourage documentation, and the front-facing camera feature encourages self-portraiture (selfies).

Along with *requests*, features that *encourage* offer the clearest depiction of designers' intentions—what designers aim for the object to do. In some cases, however, an object may *encourage* lines of action about which the designer gave little or no thought. That is, an object may be built to accomplish a specific task, and this intentionality is an obvious part of the user experience. Alternatively, the object may be built a certain way for one reason (such as aesthetics, efficiency, or cost-effectiveness) but harbor features that *encourage* seemingly unrelated user practices.

For example, sharing and engagement are *encouraged* through the Facebook architecture. Key features of Facebook—such as the immediate availability of "memories" and a visible one-click tool for sharing, a text box with one-click options for adding images, and automatic "tagging"—all combine to encourage users to generate content and connect with their networks. Adding content to Facebook is easy, seamless, and represents the intentions of Facebook, Inc. The more data users produce, the more valuable the platform is to advertisers who are willing to pay to create targeted ads and to data brokers willing to pay for users' information. It is thus in Facebook, Inc.'s financial interests to encourage data sharing and prolonged engagement, and the features of the social network site do just this. However, the business model of Facebook also, it turns out, encourages political influence. Granular and expansive data production coupled with microtargeted advertising and a hands-off moderation policy combine to create the conditions by which political operatives can construct and deliver compelling political messages to exactly those individuals most likely to be persuaded, regardless of these messages' veracity. It is unlikely that Facebook meant for this outcome, yet its product encourages the outcome nonetheless.

Dinner plates offer a less politically charged example of encouragement outside of intention. Large plates *encourage* greater food consumption, and small plates encourage portion control. Those who design and distribute dishware need not have a particular interest in consumption habits, yet plate size *encourages* and *discourages* consumption in meaningful ways. This bears out empirically, with research demonstrating that diners who eat from small plates feel more satisfied with less food than diners eating from larger plates, who require more food to feel sated.[10]

In most cases, dish design is a function of aesthetic style and normative cultural convention rather than concern about consumers' dietary practices. For instance, fine dining establishments may select large plates to enhance presentation rather than to serve large meals, and small plates may derive from normative conventions of tea settings as part of a cultural food practice rather than a portion-control strategy. However, despite designers' indifference to diet, plate size nonetheless *encourage*s more or less food consumption.

This is not to say that plate proportions, as a feature, cannot contain volume-related intentions. For example, there is an emergent market of dishware designed specifically for dietary practice. Capitalizing on the affordances of plate shape and size, companies have created food-management plating that controls portions and also *encourages* balanced food consumption. For example, the Portions Master Skinny Plate offers presized cut-outs for protein, starch, and vegetables. As described on the Portions Master website:

> The Portions Master is a portion control plate that was specifically designed to help you eat healthy and lose weight, without having to count calories. With Portions Master, you just portion out your protein, complex carbohydrates, and fiber in the appropriate space, remove Portions Master from your dinner plate, and you're ready. It's really that easy![11]

If a person wishes to eat more healthfully, the Skinny Plate accommodates. If a person wants to indulge, a larger plate without cut-out portions would pose the fewest barriers and *encourage* unrestricted consumption.

In short, technological objects *encourage* particular lines of action by making them easy and accessible. Should users wish to engage those lines of action, the object readily abides. In some cases, like Facebook's encouragement to share and Portions Master's encouragement to eat a balanced diet, the design reflects a clear set of intentions by which design collaborates with the user to coalesce in a predictable and intended outcome. However, sometimes objects encourage behaviors that may not coincide with designers' intentions. Such latent effects can encourage lines of action that generate surprising and unexpected results.

Discourage

Objects *discourage* when their architectures and normative structures erect obstacles. Whatever is discouraged is nonobvious and requires a degree of extra effort on the part of users. The action is available and plausible, but getting to it is not seamless. Users may need to employ creativity and technical savvy and be willing and able to circumvent norms and rules. Actions that an object discourages may be intentional or unintentional on the part of designers. Features might have been built to avoid a particular line of action, or certain functionalities might never have been considered in the design process and thus never incorporated into an apparatus.

Twitter's *discouragement* of long-form content, for example, erects obstacles by design. The Twitter platform makes space in each tweet for a specific number of characters (originally 140 and expanded to 280 in 2017), but there are ways for users to get around these character limits. For instance, users can take a screenshot of a long snippet of text and attach an image of the screenshot to a tweet. Users also can engage the platform's thread function to create a "tweetstorm"—a connected series of posts that generates a narrative. Despite these workarounds, users are tied to text limits by default and must undertake additional steps to practice verbosity.

On Instagram, users are *discouraged* from posting frequently. This discouragement is not a function of any design feature but reflects the norms of platform participants. While conducting interviews for one of my own previous studies, a participant recounted a story in which her younger sister was appalled to learn that the participant had posted twice within a few hours. The younger sister explained that there was a firm one-post-per-day rule and that anything beyond this was "clogging the feed."[12] The design features of Instagram do nothing to limit documentation and sharing (in fact, we might argue that the platform design *requests* and *encourages* content production and distribution), but the community informally censures those who share too much, thus discouraging abundance and enforcing relative scarcity.

Combining community norms with design features, platforms and forums that curate through voting ostensibly *discourage* dissenting voices. Designers likely implement voting features to foster democratic participation. In practice, however, voting amplifies voices from the center while minimizing or erasing voices from the margins. Those who engage in ways that resonate with the majority of the community will receive positive feedback ("upvotes") and be rewarded with increased opportunities for attention. Those who engage in ways that challenge the community will receive negative feedback ("downvotes") or be ignored.[13]

For example, the image-sharing site Imgur sorts content by "up" and "down" votes from within the community. Images and comments with the most "upvotes" are located at the top of each page, and those that receive enough "upvotes" appear on the "front page," optimizing visibility. In contrast, "downvoted" content gets pushed to the bottom of the page, and when votes go negative (receive more negative than positive votes), the content disappears from the main site. After content disappears due to a negative vote score, it remains accessible only behind a "bad comments" link. In practice, this means that users who express alternative opinions are given less space on the platform than those who express popular views, thus reinforcing the ideological status quo among community members.

In a study of engagement around racial imagery on the Imgur platform, sociologist Christopher M. Julien found that the general zeitgeist on Imgur is one of colorblindness and "postracial" humor.[14] Julien's study showed that forum participants downvote both explicit racism and progressive antiracist discourse.

Imgur's user base, which is predominately white, male, and middle class, effectively perpetuates a comfortable racial ideology and discursive practice that both rejects extreme white supremacy and also dismisses those who point out continued patterns of racial oppression. The vote feature thus *discourages* dissenting voices, empowering the community to remain ideologically unchallenged. Moreover, if we assume that dissenting voices are more likely to come from users who do not share the white, male, middle-class demographic, Imgur's vote feature also discourages participation by diverse and marginalized groups.

To be sure, dissent is possible on Imgur (and other vote-curated platforms like Reddit and the late YikYak), and there are no direct technical forms of racial or gender exclusion from participation. We may therefore imagine marginalized groups converging to generate a critical mass that changes the conversation through organized voting campaigns. However, this rebellious option is effortful and deliberate. By default, vote-curated platforms reinforce like-minded thinking and perpetuate the status quo. Dissent is not precluded, but it is socially and technologically *discouraged*.

Refuse

A line of action is *refused* when it is implausible and/or impossible. A technological object may be designed in a way that renders certain functions untenable. That is, it may be obvious from the design that particular functionality is prohibited. For instance, a traditional cell phone refuses internet access, and users probably would never consider using a classic Nokia to browse the web. Alternatively, Objects may present the possibility of functionality but then refuse when a user attempts to enact the function. For example, someone might attempt to touch a computer screen to manipulate the display, but if the display screen cannot serve as an input device, then the content will remain undisturbed.

Sometimes, objects *refuse* as a feature of design, as when an action is intentionally and explicitly prohibited. Other times, refusals are incidental, as when a feature is unreflexively omitted during construction. For example, paywalls on digital academic journal articles (discussed previously) refuse access to those who do not pay or do not carry the proper institutional affiliation. Paywalls are built with the intention of controlling access and are a feature of the publisher's business model. In contrast, some publishers do not include direct hyperlinks between citations on a reference page and the sources referenced therein. By omitting hyperlinks, those who designed the interface prevent readers from finding referenced texts, but this was likely not an explicit consideration.

Previously, I used the example of Twitter *discouraging* long-form narratives by limiting character counts. Here, we may say that Twitter *refuses* to accommodate more than 280 characters in a single communication. This refusal is a feature of the interface design. When a user exceeds the designated character limit, the excess text turns red, as does the "remaining characters" indicator at the bottom of the tweet. The "remaining characters" indicator also displays negative numbers, showing users how many words beyond the designated boundary they have typed. The Post button fades to pale blue and becomes inactive, thus refusing a communication that is over the limit. In this way, Twitter refuses more than 280 characters per tweet, thus *requesting* brevity.

A key feature that distinguishes Facebook from its predecessor MySpace is that the latter *encourages* page personalization while the former *refuses* personalization. Facebook users can provide content exclusively within prefigured categories set by the platform. The platform refuses to add music or background designs to a user's profile. The prefigured categories are arranged in a set order and displayed in a uniform way for all users. Facebook also refuses to let users publicly rank order their friends, a feature integral to the MySpace architecture.

Recalling Robert Moses's bridges in New York City, we can say that low overhangs *refused* to let public transport buses through. Moses's urban planning design did not refuse access to economically disadvantaged people but *discouraged* access by restricting forms of transit on which less wealthy New Yorkers were more likely to rely. Scholars have debated about Moses's racist or classist intentions when designing the bridges,[15] but intentional or not, the bridges continue to organize movements of people, cars, real estate, and money along lines of race and class.

Objects *refuse* by excluding and prohibiting specific acts. These may be integral to the design or unreflexive products of omission. Indeed, the emergence of new features often transform an object that once refused into one that *encourages* (or vice versa). For instance, early cell phones were not fitted with cameras, thus refusing pictoral documentation. Newer models, however, have cameras and applications for photo storage and sharing, encouraging users to snap pictures and to do so in a social manner.[16]

It again bears reminding that *refuse* (like *demand*) is not a deterministic category. Refusals are not necessarily universal, nor are they always permanent. They present as impossible but remain subject to change and circumvention. Twitter previously refused more than 140 characters but now enables up to 280.

Facebook users may populate their photo streams with a particular aesthetic to approximate background personalization, even though "wallpapers" as such are refused by the platform.

In sum, objects *encourage* by making some lines of action obvious and easy, *discourage* by making some lines of action difficult to access, and *refuse* by rendering some lines of action impossible or implausible. These categories are not fixed nor are they mutually exclusive. For instance, leaving a tip is gently encouraged by the presence of a tip jar on a countertop but more explicitly encouraged by including a preset tip option as part of the card payment process. Shoppers are subtly discouraged from printing a receipt when an electronic self-checkout prompts the customer to make a decision (by asking, "Would you like to print a receipt?"), more strongly discouraged when the default response is no (leaving the shopper to switch the selection to yes), and even further discouraged—and potentially refused—if a receipt materializes only after the shopper asks for a copy from a human cashier.

Allow

Technologies place bids on users in the form of *requests* and *demands*. Technologies respond to users in the form of *encouragement*, *discouragement*, and *refusal*. *Allow* applies to bids placed by technologies and to bids placed on technologies. Allow is distinct from other mechanisms of affordance due to its neutral intensity and multidirectional application. A user may take a line of action, but there is no pressure to do so, and there are no significant obstacles in the way. Allow is like a fork in the road. A traveler may just as easily opt for one route as another. The traveler is not faced with enticements from any direction, and the traveler does not need to overcome any extra blockades to access the pathways.

For example, multispeed blenders and multilevel light dimmers *allow* people to select variants of power and brightness levels at their own discretion. The blender does not try to persuade the cook to pulverize rather than fold, and it obliges equally when the cook pulses and continuously churns. The light dimmer does not resist the slightest glimmer or the most brilliant glow but allows light dispersion as the user deems fit.

Although Twitter, Instagram, and Snapchat *request* that users share content and *encourage* users to connect with others on the platforms, all of these platforms *allow* users to select any username handle that they wish. This naming policy is distinct from Facebook's, which, through the company's terms of service, has always *demanded* that users display their "real" names. However, after much controversy and debate, Facebook now allows users to select a "real" name from their name assigned at birth or another name by which friends and family would recognize them.

Just because *allow* is neutral in tone does not mean it is apolitical. For instance, the Facebook name policy has been mired in political contention, with opponents pointing to privacy concerns, especially among marginalized populations who might find harm in identity exposure. The allowed detachment between user identities and handles on Twitter, Instagram, and Snapchat thus accounts for issues of privacy and attends to a range of user vulnerabilities in ways that the Facebook platform neglects. Allowing open user handles is therefore a political decision.

Sometimes, features that maintain indifference by design are deeply political in ways entirely unforeseen and unintended by designers. Remember, when left unchecked, technologies will arc toward power and privilege. This point became clear in the exposure of ostensibly neutral advertising interfaces of major digital media platforms during racial unrest in the United States in 2016 and 2017. As of this writing, the advertising interfaces for leading digital media platforms *allow* customers to utilize granular data to target any group of interest. After the appearance of a disturbing amount of white supremacist propaganda during the 2016 US presidential campaigns, journalists at Pro-Publica entered the Facebook ad interface to investigate the capacity to target users with white nationalist leanings. The publication identified 2,300 users who had expressed interest in "Jew hater," "how to burn Jews," and "history of 'why Jews ruin the world.'" With a quick fifteen-minute approval process, ProPublica was able to "promote" content to these anti-Semitic targets.[17] Journalists at BuzzFeed similarly tested how Google's ad interface handled racist inputs. The BuzzFeed team typed the keywords "white people ruin," and the ad platform suggested running advertisements next to searches for "black people ruin neighborhoods." With the keywords "why do Jews ruin everything," Google suggested ads tied to searches for "evil Jews" and "Jewish control of banks."[18] Similar issues were found in the ad interfaces of Twitter and Instagram. In short, building algorithms that *allow* targeting from any direction and with any agenda is a political decision because it forgoes an alternate option that *refuses* racism and expressions of hate.

Although Facebook, Google, Instagram, and Twitter all prohibit racism and bigotry in their terms of service, the designs of their platforms do little to uphold these rules. Their algorithms are built to extract data with optimal granularity and to churn that data into information for paying customers. These companies project values of inclusion and equality, but their platform architectures do not have these values encoded. Thus, *allow* is neutral in intensity but can maintain strong political leanings, connected to or separate from, designer intent.

Chapter Summary

Despite Gibson's and Norman's articulations of affordances as gradated and nuanced, applications of the concept have been persistently static and binary. The struggle to incorporate gradation stems from a flawed analytic starting point. As long as analysts begin by asking, "What do these objects afford?," they will remain stuck in imprecise formulations by which an affordance is either present or absent. Altering the question to "How do these objects afford?" creates space for dynamism.

Asking *how* instead of *what* lays the groundwork for developing a framework and vocabulary that captures the continuous nature of technological affordances. Beginning with how, I suggest that affordances are characterized by a suite of mechanisms: *request*, *demand*, *encourage*, *discourage*, *refuse*, and *allow*. These mechanisms operate together as a tool to discern and articulate the varying degrees of insistence with which technological objects push, pull, and respond in multiple directions.

Rather than a rigid framework, the mechanisms are porous, rendering the analytic tool malleable by design. Each mechanism is an artificial stopping point rather than a firm designation, and within each mechanism, there remains room for variation. *Demands* may be strong or weak, resisting or seeping into the borders of *requests*. It may be unclear whether an artifact *refuses* some action or just firmly *discourages* it. A feature may sit ambiguously between *encourage* and *allow*. Indeed, one might say that as a tool, the mechanisms and conditions framework encourages and requests disagreement and debate in affordance analyses.

The implications of the mechanisms of affordance can be individual, interpersonal, and/or cultural-structural. As features push and pull with varying degrees of insistence, these features guide what people do, how they interact, and how macro-level patterns are formed, altered, and reified. For instance, swipe-based dating apps *request* that users consider a high volume of potential partners and *discourage* users from slow considerations. The swipe feature may then shape how individuals evaluate potential partners and how they present themselves as romantically appealing—placing emphasis on quickly identifiable markers such as physical attractiveness and income. The glut of potential partners and ease of selection and dismissal may shape how those who use the apps interact during dates, perhaps moving quickly to intimacy to establish commitment within a crowded pool or keeping distance to avoid foreclosing the full range of romantic options. These micro interactions can affect romance and intimacy at a cultural-structural level by normalizing serial dating, detaching a single date from future romantic engagements, and empowering those who feel dissatisfied in current relationships to explore the abundant field. In short, swipe apps don't just offer another way to date but reshape the meaning and practice of finding love.

The mechanisms of affordance hold social, political, economic, and legal ramifications, with far-reaching effects. Higdon's HB37 in Missouri, for example, was about more than just controlling space or enforcing safety. The timing of the bill coincided with US protests about racist policing practices, including violence by police officers against black citizens. Some of the most tumultuous protests took place in Ferguson, Missouri—the state in which HB37 was introduced. Attempts to criminalize police barrier breaches can therefore be read as a political move that restricts protest activities and shifts power to state authorities. HB37 would not only strengthen the *request* that citizens remain outside of police perimeters but also *discourage* aggressive protest tactics and *encourage* police use of force.

In sum, the mechanisms of affordance address the binary problem that has heretofore plagued affordance theory. Asking *how* instead of *what* gives nuance and agility to affordance analysis, freeing it from rigid binary constrictions. However, the mechanisms alone are not enough. On their own, the mechanisms of affordance depict complex objects in relation with homogeneous subjects. But affordances will vary across users and contexts. Thus, we must ask not only *how* objects afford but also *for whom and under what circumstances?*. This question—*for whom and under what circumstances?*—is the focus of the next chapter.

5 Conditions of Affordance

In 2014, a fifty-seven-year-old woman was arrested and charged with "interfering with a peace officer" after crossing police tape during a standoff between law enforcement and a potentially violent suspect near her home in Eugene, Oregon. As discussed in the previous chapter, police tape offers a flimsy physical barrier, but symbolically (and legally), it maintains sway. An article about the incident in Eugene's *The Register-Guard* quotes officers expressing concern about the integrity of the scene and the safety hazard of barrier breaches.[1] Officers also describe the woman's behavior as irresponsible and insubordinate, noting that she was smoking cigarettes and appeared intoxicated. Said police spokeswoman Melinda McLaughlin:

> People may be curious, but these are high risk situations and there is a reason why there is a perimeter. . . . They warned her a few times, but she kept coming out to smoke and asking what was going on. . . . It was taking officer resources to manage her. . . . She was intentionally failing to follow commands.

McLaughlin's justification seems reasonable. Removal was about safety, the police barrier is a legal perimeter, and this woman was acting unruly. We might expect that such measures would be taken against any person who similarly transgressed. However, a line toward the end of *The Register-Guard* article stands out as curious:

> A second person—a member of the media—also crossed the perimeter, McLaughlin said, but left before officers issued a warning.

Two people traversed a legal perimeter on the same night, in the same place, during the course of the same police operation. One was arrested. The other was never even warned—though the other's presence was noticed, as indicated by McLaughlin's statement.

The apartments in which the incident occurred are located in Eugene's Jefferson Westside neighborhood. This neighborhood has a median income under $30,000 (47 percent below the national average) and a crime rate that is 61 percent above the national average.[2] As a resident of this area, living in a modest rental unit, the arrested woman did not benefit from valued social class signifiers. Rather, her resident status near the crime scene likely undermined the legitimacy of her presence and reinforced the police boundary via heightened vigilance and mistrust. In other words, the police were disinclined give this resident the same benefit of the doubt granted to a journalist who walked through the perimeter noticed but undisturbed.

Far from objective, the surveillance, suspicion, and punishment around police tape took shape in distinct ways for two different subjects. For the media associate, the perimeter proved porous; for the resident, the tape was iron clad. Notably, images from *The Register-Guard* article indicate that the arrested woman racially presents as white. Given existing statistics and accounts about police interactions with racial minorities, especially those of low socioeconomic means, we might imagine that the barrier formed by the police tape would have been even stronger—and the consequences more severe—had the resident been a person of color.

The point is that the mechanisms of affordance—*how* objects afford—are necessarily entangled with social and structural conditions. Affordance analyses thus begin with the two-part question: *How does this object afford* and *for whom and under what circumstances?*. From this analytic base, affordances are neither singular nor static but protean relationships between artifacts, persons, and situations that remain always, potentially, in flux. *For whom and under what circumstances?* is represented in the framework by the "conditions of affordance."

In this chapter, the conditions of affordance are distilled into three broad factors: *perception*, *dexterity*, and *cultural and institutional legitimacy*. These factors address how subjects perceive objects and the functionality, barriers, opportunities, and constraints therein; the skill with which subjects can engage objects; and the degree to which the subject-object relationship is sanctioned by normative conventions and official codifications (that is, norms, rules, and laws). As stated in the framework's original formulation:

> evaluating an artifact's affordances entails discerning if a subject perceives the artifact's function, and if so, does that subject have the physical and cognitive dexterity to utilize it, and if so, is the subject's use of the artifact culturally valid and institutionally supported.[3]

In simpler terms, what an object *demands* of me may be only a *request* for you. I may be *encouraged* in some instances and *refused* in others. You may be *allowed* to enact some function, but I may not. Affordances are built into material features but only partially so. Proper affordance analysis requires attention to features in context.

The three broad conditions of affordance are not discrete categories but are entwined in mutually shaping relation. Each factor informs and is informed by the others. *Perception* is likely affected by the skill or *dexterity* one has with an object, just as *perception* of the object can enhance or hinder user competence. Clear *perception* and skillful *dexterity* may earn a subject *cultural and institutional legitimacy*, just as *cultural and institutional legitimacy* may foster skill development.

As typological demarcations, *perception*, *dexterity*, and *cultural and institutional legitimacy* echo existing work from affordance theory and from the practical turn in design studies. In a conceptual review of affordances, Joanna McGrenere and Wayne Ho identify two axes along which users may experience variation in the affordances of an object: "the ease with which an affordance can be undertaken and . . . the clarity of the information that describes the existing affordance."[4] From a design perspective, Batya Friedman and David G. Hendry argue that critical and reflexive design should account for variations in cognitive, technical, and physical competency.[5] Perception, dexterity, and cultural and institutional legitimacy operationalize such variations into a usable model.

The conditions of affordance not only add context to analyses but also reveal the default subjects for whom technologies are designed. Identifying who is *refused* versus who is *allowed* or *encouraged* to access technological features clarifies a set of assumptions about imagined users—their social positions, physical characteristics, and material and immaterial resources. In turn, these identifications also render visible those who are marginalized, ignored, and excluded. This gives depth and breadth to analytic understandings of human-technology relations. It should also give pause to practitioners (including designers, engineers, executives, and investors) whose products are rooted in and distributed through complex social systems.

Perception

In an update to his 1988 *The Psychology of Everyday Things*—renamed *The Design of Everyday Things*—Donald A. Norman makes a distinction between real affordances and perceived affordances.[6] Real affordances refer to the material features of an object, and perceived affordances refer to the way subjects interpret those features. How an object affords thus depends partially on the extent to which a subject is aware of an object's functionality. Without subjective awareness, the features of an object remain inert.

The inclusion of a technical function is a necessary but insufficient condition for its availability. A function about which a subject is unaware is as effective as a function that is absent. This is exemplified in an observation by communication scholar Gina Neff, who points out that "For hackers and experts, systems look more crackable, more full of potential and possibility, than they look to the rest of us—appearing to us as given and relatively fixed."[7] Thus, the material elements of a technical object emerge not objectively but always through a lens.

Imagine coming home from work at the end of a long week, looking forward to a quiet glass of wine. You walk in the door, toss your coat, and uncork a bottle. Opening your cupboards, you see that there isn't a single clean glass in the house. You do not feel like washing dishes. After a brief pause you realize that although you may not have clean glasses, you do have measuring cups. "That'll do!" you think, and pour yourself a drink.

Glasses and measuring cups are designed for different purposes. The former is meant to hold beverages, and the latter is meant to portion food and cooking materials. One is for consumption; the other, for preparation. Yet many features of drinking glasses and measuring cups are largely interchangeable. Both can hold consumable substances and enable human subjects to transfer those substances from one vessel to another, whether into a bowl or into a mouth. For the measuring cup to become a drinking glass, however, the subject must perceive this functionality. That is, the measuring cup is a drinking glass only after it is recognized as such. Without recognizing the measuring cup as a potential drinking device, the cup *refuses* direct consumption. In contrast, once a subject perceives the measuring cup as having glasslike features, that subject is *allowed* to drink directly from it.

On social media platforms, algorithms actively curate both content and relationships. Some content and people are highlighted, and others are relegated to the bottom of the feed or entirely omitted from view.[8] One of the main critiques leveled against digital social platforms is that users often do not—and cannot—understand how curatorial decisions are made.[9] *Perception* has thus emerged as a critical issue around truth and trustworthiness. By default, the features of most social media platforms functionally *discourage* or even *refuse* critical investigation into the source of content and its place within an information stream. Hence, crises of "fake news" and misinformation have occupied public attention and prompted a flurry of responses from social media companies, which have sought to institute means of truth verification via technical design and personnel.

(Mis)information does not affect everyone equally. Novices might not know that their news is presorted or that they can readjust the flow of content. In contrast, those equipped with a higher level of media literacy are less bound by default algorithms, can perceive their news feeds as both constructed and pliable, and can alter the feeds to fit personal information preferences. For media-savvy subjects, default settings are *requests* rather than *demands*, and the information that does filter through *allows* for skepticism. Such tactics might defend against dubious information, *encouraging* critical news consumption. At the same time, adjusting the default options can just as easily exacerbate the problem of misinformation because those familiar with platform settings are *allowed* (or even *encouraged*) to filter out content that challenges their personal worldviews, reinforcing tight filter bubbles that *request* confirmation biases and *discourage* encounters with opposing perspectives. Perception thus affects the fixed (or pliable) nature of a news feed but does so toward multiple, sometimes contradictory ends.

Perception is not always liberating or enabling. In some cases, clear perception of features—in their material and social forms—can have constraining effects. In the previous chapter, one of my interviewees was reprimanded by her younger sister after sharing "too many" posts on Instagram. These two sisters did not follow the same normative conventions. Thus, my participant was initially *allowed* to post as many times as she wanted, while Instagram strongly *discouraged* the same actions from the participant's sibling. Indeed, the experience of censure altered my participant's perception such that the affordances of Instagram changed for her. What was once *allowed* was no longer.

Perception has been and remains a crucial variable in the conceptual trajectory of affordances. It is the crux of ontological debates about affordances as intrinsic properties versus relational elements—that is, is the existence of an affordance inherent, or do features afford only once they are perceived? This question—akin to those about the sounds of trees falling in forests—is a philosophical one. In contrast, the mechanisms and conditions framework is a practical project, and I am less concerned about the nature of affordances than how they operate. In practical operation, perception activates affordances and alters their shape, rendering technologies more flexible or more constricting. Perception shifts subject-object relations between various mechanisms from *request* to *refuse*, accommodating and subverting the intentions of design.

Dexterity

In order to utilize the features of an object, one must not only perceive those features as available but also have the ability to employ them. *Dexterity* refers to the capacity of a subject to enact the functions of an object. These capacities can be physical or cognitive. A subject must be able to physically manipulate the object in required ways and have the knowledge set that enables manipulation.

Variation in *dexterity* and its relationship to affordances is a central element of disability studies and disability activism. By and large, built infrastructures have historically been constructed with a presumed model human who walks easily, sees clearly, and hears with precision. The critical disability perspective contends that such assumptions have resulted in a disabling social structure for those whose bodies do not adhere to the presumed model ideal.[10] Stairs offer a clear example. Stairs are built to transport bodies between levels of a given built structure. In the default case, stairs *encourage* climbing. However, as described in William H. Warren's classic affordance study, stair climbing relies on particular bodily configurations.[11] In addition to adequate leg-length-to-stair-height ratios, stair climbers must contain the muscular capacity to support full body weight, the muscle control to contort the legs at will, and the coordination to balance for brief moments as one leg leaves a bottom step and advances to the next. Persons with lower body paralysis, muscular atrophy, severe arthritis, or myriad other conditions do not have bodies that adhere to these requirements. Thus, although the stairs *encourage* climbing by able-bodied persons, those living with certain mobility impairments are *discouraged* or *refused*.

In this vein, visually oriented websites serve consumers who have clear ocular vision, but they *discourage* or even *refuse* access among people with vision impairments. For instance, a stylish thin font presents a clean look but may be unreadable for those with eyesight less than 20/20; screen readers translate website content into audio form but cannot read images directly and often have trouble reading text within images; and similar color contrasts (such as white on grey) can render text indistinguishable from the background. So although website designers work hard to create slick visuals, the process of doing so may *request* engagement from seeing users while *refusing* access to the visually impaired. In contrast, websites that adhere to the World Wide Web Consortium's (W3C) standards of accessibility[12] equally *allow* and even *encourage* consumption by users with any level of vision.

On a personal note, *dexterity* plays a substantial role in my collaborative research relationships. Like all researchers, I operate with clear limits. I am highly proficient in qualitative methods and have a strong grasp of social theory. Yet my dexterity with both statistical analyses and large-scale data analytics is relatively basic. I do not have the skills to build complex statistical models or write novel code that unlocks and makes sense of data points from the web. On my own, I am *allowed* to use a variety of data collection and analysis tools but *refused* all but their most elementary functions. Luckily, I work with wonderful coauthors who do have expertise in statistics and big-data digital methods. Because of my coauthors' dexterity in this regard, the features of quantitative and digital data analysis software *allow*, *request*, and *encourage* complex analyses for our research teams.

Dexterity, like *perception*, is not a fixed designation but a description of subject-object relations at a particular juncture. Prohibitions need not apply forever, nor is access guaranteed over time and across contexts. This is because competencies can change, and so too can circumstances. For example, I could upskill via research methods workshops and classes; a person with adult-onset vision impairment might initially find screen readers unintelligible but with time and practice could become adept; and the features of a new phone may seem inaccessible at first but quickly become familiar, *allowing* and *encouraging* use of various features toward multiple ends. In turn, one's *dexterity* with a set of stairs may decline with age, and an operating system update might upend one's previously intuitive and expert relation to a device.

Cultural and Institutional Legitimacy

Growing up in my parents' home, I was aware of how the thermostat worked. I knew what it did and how to operate its functions. However, in my eighteen years of living at home and subsequent visits throughout adulthood, I have yet to change the temperature. My structural position within the family—coupled with the practical fact that I don't pay the bills—creates a circumstance in which I am strongly *discouraged* from operating temperature control technologies in the family dwelling. I am *allowed* to use a space heater or a fan and *encouraged* to use tools like sweaters and socks to control my personal temperature, but heavy barriers remain in place that restrict control over central temperature technologies. In fact, my family has very particular conventions about who may access temperature controls and under what circumstances. My mother has control in the summer, and my father has control in the winter. My brother and I should never touch the thermostat. Thus, my mother is *encouraged* to deploy temperature controls in the warmer months and *discouraged* but still *allowed* to do so after October. My father is *discouraged* in the winter but *encouraged* in the summer. We "kids" face heavy *discouragement*, if not outright *refusal*, year-round.

Sociotechnical assemblages exist at the intersection of history, biography, and culture. Cultural norms and institutional codes reflect and shape social and political dynamics, and these dynamics inform the way people and technologies relate. Thus, the force exerted by technologies is inextricable from the structural position of social subjects. As a condition of affordance, *cultural and institutional legitimacy* addresses the way one's location within the larger social structure and the related norms, values, rules, and laws of a social system inform human-technology relations.

Cultural and institutional legitimacy can operate through both formal and informal channels, representing a continuum between codes and conventions. In some cases, the affordances of a technology are tied to formal rules and laws (codes). In other cases, affordances are guided by normative patterns (conventions). Recalling the police tape, this technology *allows* and *encourages* breach by police officers but not citizens. Citizens are legally *refused* access, but police officers are authorized to pass. Between citizens, the pressure exerted by the police tape varies less by code and more by norms or conventions. In the opening to this chapter, I recount a story of a local resident who was arrested after breaching the police barrier during a standoff while a journalist entered and exited without warning or censure. The journalist was marked by signifiers of high status that buffered against suspicion and surveillance, in contrast with the local woman, who quickly became suspect. The barrier was thus a *refusal* for the local citizen but a mere *discouragement* for the journalist.

Cultural and institutional legitimacy is an intrinsically political condition tied to existing status and power dynamics. How access is distributed and for whom technologies are (implicitly and explicitly) intended reflect status markers within the broader social system, most often privileging those with valued status traits. By default, technologies will lean toward the reification of power structures. Subverting power structures thus requires attention to the ways that existing sociotechnical systems serve, ignore, or harm, socially situated subjects.

In a study of flagging and reporting on social media, Stefanie Duguay, Jean Burgess, and Nicolas Suzor examine how platform features uniquely affect queer*-identifying women.[13] Their analysis of Instagram, Tinder, and Vine documents architectural elements, terms of service, and normative community practices. They show that content moderation features across these platforms serve a default user and that the default user is not queer* women. For example, Instagram has a highly visible reporting mechanism but moderates based on heteronormative conventions, Tinder has an obscure flagging feature that enables sexually exploitative and deceptive behavior,[14] and Vine's laissez-faire approach implicitly supports a "toxic technoculture" that allows antiqueer* sentiments to proliferate. Cultural norms and interface design thus construct inhospitable environments for LGBTQI women. Although the platforms *encourage* participation among one demographic (cis white people), they *discourage* participation for those who fall outside of this imagined demographic ideal.

Another interesting example comes from the literature on social media and social capital. It has emerged axiomatic that social media use enhances social capital in the form of network building, information sharing, and resource distribution. The truism of a positive relationship between social media and social capital derives primarily from studies of university students.[15] Although the authors of these studies are clear about the applicable scope of their findings (university students) and although social media has also proven capital-enhancing in other populations (such as the elderly),[16] the idea that social media is *universally* capital-enhancing has morphed into a general empirical claim. However, subsequent work shows that social media experiences are highly variable and for some can be more of a liability than an asset.

In sharp contrast to university-based findings, ethnographic exploration of social media experiences among low-income urban youth of color paint an image of vulnerability rather than opportunity.[17] Here, social media are not sites of resource accumulation but surveillance, drama, and danger. These works report on youth who face threats of violence, have private images exposed, and see fights escalate, leading to self-imposed restrictions on sharing and participation. Far from the happy "highlight reels" that characterize social media experiences within the public imagination, these urban youth are *discouraged* from the network-building features of social media platforms, are *refused* the freedom to connect, and face *requests* for limited engagement. That is, social disadvantage is not ameliorated by connectivity but is compounded as youth move online.

In short, human-technology relations and the affordances therein are always socially situated. Cultural norms and institutional codes create both opportunities and prohibitions. Cultural and institutional factors do not determine human-technology relations, but pathways are built with wider or narrower entrances and with smoother or rockier terrains. The contours of these pathways create opportunity structures that shape sociotechnical dynamics at both micro and macro levels.

Chapter Summary

This chapter addresses the second part of the mechanisms and conditions framework, demonstrating that affordances take shape in relation to diverse subjects operating under a range of contextual variables. Discerning *how* objects afford (the mechanisms) entails careful analysis of *for whom and under what circumstances?* (the conditions). The conditions of affordance are encoded into built objects but necessarily go beyond materiality. How objects afford will vary from one person to the next, from one circumstance to another, and in new ways over time.

The larger message of including "conditions" in the mechanisms and conditions framework is that humans and technologies are co-constitutive and structurally situated. Politically aware analyses necessitate careful consideration of the material and the social. The conditions of affordance facilitate analyses in which the same object affords in multiple and divergent ways. Indeed, the conditions of affordance insist that technical features are polysemic, polyvalent, and variable. In its specific articulation, the conditions of affordance include three factors: *perception*, *dexterity*, and *cultural and institutional legitimacy*. Although analytically distinct, each condition entwines with and informs the others.

In some cases, a person may recognize what a feature does (*perception*), and have the skill to use it (*dexterity*), but face formal or normative barriers that *refuse* or *discourage* enactment. For example, a young hacker may know how to access and change school records but is prohibited from doing so by threat of expulsion or legal action. The refusal persists despite perceptive awareness and practical capability. In other cases, *cultural and institutional legitimacy* can facilitate or impede familiarity with an object, thus affecting perception and dexterity. For example, men have been cast as more technologically inclined than women, creating a relationship of exploration and technical skill-building among boys and making this path less appealing or obvious to girls. If technological competence is accepted and expected, perception and dexterity more easily follow; if technological competence is deterred or not assumed, perception and dexterity are less likely to develop. Further still, perception and dexterity can be the basis for cultural and institutional legitimacy. For instance, the institutional legitimacy to operate a motor vehicle (a driver's license) is predicated on demonstrating familiarity with vehicular operations and adeptness behind the wheel.

The conditions of affordance are both reflective and productive. They embody existing sociostructural arrangements and reverberate out to shape cultural norms, institutional practices, and the practical realities of everyday life. For instance, in regions where it is illegal for women to drive, these laws shape and reflect more than women's relationships to motor vehicles. They also shape and reflect gendered relations of dependence, restrict women's employment opportunities, and symbolically entrench a clear gender-status structure in which men maintain disproportionate power in the home and in society. In this way, students who lack access to personal computers in the home may be less adept at operating the features of digital technologies, including navigation of online learning platforms. Features of online learning platforms are therefore more readily available to "highly connected" learners, constructing a relationship in which digital inequality comes to affect educational outcomes that, in turn, shape life chances in ways that reproduce wealth and poverty.

The conditions of affordance take shape not only through direct interactions between individual subjects and individual objects but also through multifaceted sociotechnical relationships. The affordances of some object for some person in some circumstance can change with the introduction of a complementary technology or the inclusion of other people. Screen readers, for instance, combine with existing website architectures to make content legible to the visually impaired, shifting the affordances of the site from *refuse* to *allow*, *request*, and *encourage*. Even without an electronic screen reader, the issue of accessibility is not insurmountable. The features of a website can also become available through collaboration with another person who may read the site aloud (an organic screen reader). The inaccessible website may not *encourage* use by people with vision impairments, but with the proper tools or collaborations, consumption is *allowed*. My own research partnerships highlight the relationality of sociotechnical assemblages. The features of some data analytics software *discourage* me from using them due to my low levels of *dexterity*, but I am welcome to mobilize the features with the aid of skilled colleagues.

The conditions of affordance are not fixed to individual persons or contexts. Circumstances change, and when they do, so too do the relationships between human subjects and technological objects. In 2017, Mohammed bin Salman of Saudi Arabia issued a decree that legalized driving for women in the country.[18] This legal decree altered the cultural and institutional support available to Saudi women who wished to operate vehicles. Many of the car's affordances were previously *refused* to women by threat of imprisonment but now would be *allowed*. For some women, this legal change *requested* and *encouraged* driving (though a lifetime of being a passenger might inhibit *perception* and *dexterity*, posing barriers to use). For women who continue to adhere to traditional beliefs and are embedded in traditional networks, operating a motor vehicle is, at best, *allowed*. Indeed, the normative cultural milieu of this latter group may still *demand* driving abstinence.

At a meta level, the conditions of affordance both situate analytic outcomes and also sharpen the analytic process. Technology commentators are necessarily entrenched in cultural systems. From a first-person perspective, affordance analyses will be colored by the social position of the analyst. Whether some feature reads most clearly as *request* versus *allow*, for example, may well be a function of the conditions under which a particular analyst encounters a particular object (such as cultural and institutional access to that object). The conditions of affordance thus not only contextualize *how* artifacts afford but also *encourage* analytic reflexivity. If analysts must always ask *for whom and under what circumstances?*, they foreground their own default assumptions about the opportunities and constraints of the technology under consideration. The conditions of affordance thus reveal potential biases and infuse the analytic process with critical self-reflection.

Having operationalized a model of affordances in which mechanisms and conditions combine to structure analyses of subject-object relations, we turn now to methodological strategies by which such an operationalization can be applied. What I set roots for in chapters 1, 2, and 3 and delineate in in chapters 4 and 5 is an agile and politically attuned analytic tool. The following chapter identifies and describes methodologies of implementation. Rather than suggest something entirely novel, chapter 6 instead demarcates methodological approaches that resonate with key assumptions of the mechanisms and conditions framework, thus generating symbiotic theory-methods pairings.

6   Affordances in Practice

On their own, conceptual frameworks are little better than neat party tricks, arranging complex social phenomena into tidily packaged vocabularies. The purpose of conceptual work is to help analysts and practitioners better understand and intervene in the social world. Frameworks thus enliven through empirical encounters, facilitated by rigorous methodologies. This chapter points to methodological approaches that complement the mechanisms and conditions framework and that are enhanced through the pairing. The methodologies presented herein are not exhaustive but exemplify the vehicles through which affordance analyses can take effect. Each was selected for its high quality, relevance to the fields of technology studies, and coincidence with key assumptions of the mechanisms and conditions framework.

Methodological complements to the mechanisms and conditions framework can be varied in their approaches (qualitative, quantitative, computational etc.), stem from an array of disciplines (such as communication studies, science and technology studies, engineering, and education), and serve a range of practical and intellectual goals (including accessible design, critical analysis, and policy development to name a few). Not all methods are equally appropriate, but maintaining the integrity of the mechanisms and conditions framework requires implementation through approaches that share the framework's key theoretical tenets. Methods best suited to the mechanisms and conditions framework should meet the following criteria:

1. Centralize political dynamics.
2. Give voice to marginal populations and groups.
3. Maintain a reflexive orientation.
4. Assume multiplicity of meaning, experience, and outcome.
5. Address materiality with a social lens.

Artifacts are political. They both reflect and affect social organization and the values entailed therein. Methods that assume political neutrality or ignore the political element are ill-suited for the mechanisms and conditions framework because they leave little space for critical consideration of production, implementation, distribution, and use across subjects and circumstances. Political neutrality thus undermines analytic attention to the conditions of affordance as structural factors that shape how artifacts afford. In contrast, approaches that treat politics and power as integral can mobilize the mechanisms and conditions framework to both reveal and potentially upend existing structural arrangements.

Connected to politics and power is the propensity of a given method to lift marginal voices and prioritize the non-normative. Methods that converge on the center highlight dominant perspectives and treat them as universal. In contrast, the mechanisms and conditions framework centralizes variability, otherness, and difference. This means more than simply accounting for diverse stakeholders but also intentionally amplifying those stakeholders whose experiences diverge from the "norm." Technological design (and the design of social organization more generally) trends toward normalization. That is, design often assumes a default user but does not explicate who that user is or how default assumptions foster and entrench exclusion. Methods that recognize the normalization trend and explicitly push back against it are appropriate for affordance analyses using the mechanisms and conditions framework.

Claims to objectivity are antithetical to the mechanisms and conditions framework. The mechanisms and conditions framework is flexible and reflexive by design. It is a tool with which to make arguments, not define or solidify "truths." Scientific claims are always imperfect approximations. Thus, methods that recognize the subjective nature of epistemology—or ways of knowing—are appropriate for affordance analyses. Methodological reflexivity indicates critical introspection about how research questions, data collection strategies, and analyses are based on choices that could be otherwise and if they were otherwise, would likely produce different results.

No technological object has a singular meaning, nor does it produce entirely predictable outcomes. That is, artifacts are polysemic, dynamic, and surprising. Overly rigid empirical approaches undercut the capacity for analytic and technical adaptation, while methods that embrace fluid meaning systems approach sociotechnical relations as inherently moving targets. The latter are appropriate for affordance analyses.

Finally, the mechanisms and conditions framework is best applied through approaches that imbricate the material with the social. Specifically, methods should account for the material elements of an artifact while avoiding technological determinism. Appropriate methodologies will recognize architectures, infrastructures, and features as opportunities and constraints rather than as direct causal forces.

In the remainder of this chapter, I discuss five methodological approaches that meet all of these criteria (political, marginal, reflexive, multiple, and sociomaterial). I delineate each methodology, reference key works, and think through how each approach might be used in conjunction with the mechanisms and conditions framework. Other methodologies also fit the criteria and can be effectively coupled with affordance analyses. I present these five as exemplar cases. Along with highlighting methodological tools appropriate for affordance analyses, the chapter also demonstrates the applicability of the mechanisms and conditions framework across a wide variety of empirical topics. The chapter thus highlights the breadth of the mechanisms and conditions framework while providing practical guidance for research-based application.

Critical Technocultural Discourse Analysis

Of all the approaches featured in this chapter, critical technocultural discourse analysis (CTDA) shows the tightest fit with the theoretical underpinnings of the mechanisms and conditions framework. Introduced by digital media and race scholar André Brock, CTDA specifically applies to analyses of internet technologies and entails simultaneous attention to hardware, software, ideology, and user experience.[1] CTDA combines material analysis of hardware and software design as it intersects with meaning production and articulation by socially situated users. The method entails deep and simultaneous consideration of *artifact*, *practice*, and *belief*, which together create analytic texts that the researcher reads through a critical lens.[2] CTDA centralizes the social margins and lifts those voices that are otherwise unheard, positioned against the norm, or considered only retroactively.

Key assumptions of CTDA include the following:

- ICTs [information communication technologies] are not neutral artifacts outside of society; they are shaped by the sociocultural contexts of their design and use.

- Society organizes itself through the artifacts, ideologies and discourses of ICT based technoculture.

- Technocultural discourse *must* be framed from the cultural perspective of the user AND of the designer.[3]

Brock warns that as a method, CTDA is "unwieldy" and inefficient and that "neither interface analysis nor critical discourse analysis can be done succinctly."[4] A CTDA reading entails analytic accounts of interrelated materiality, practices, and culture and focuses on power dynamics. CTDA thus insists on data that are multiply analyzed and articulated.[5]

Like the mechanisms and conditions framework, CTDA eschews positivist notions of objectivity and determinism in the research process. CTDA analysis generates not definitive answers but situated arguments. Data are always read through a lens, such that different analysts may come to different conclusions about the same research object. This resonates with the porous nature of affordance mechanisms and their use as analytic pegs rather than as concrete categorizations. Hence, a *request* may be argued as a *demand* or from another perspective formulated as a *refusal*.

For both CTDA and the mechanisms and conditions framework, ambiguity is a feature, not a bug. Both resist definitive statements and in doing so destabilize analytic authority. This does not equate to imprecision but rather, baked-in discomfort and uncertainty. Together, CTDA and the mechanisms and conditions framework create a vehicle for thoughtful, critical, and rigorously derived scholarship by which knowledge remains a living and changing organism. Conclusions remain always contentious, situated, open to critique, and subject to change.

As indicated by the name, critical technocultural discourse analysis maintains a critical perspective. Drawing on critical information studies, CTDA focuses not on *problems* but on *problematics*—issues like racism, sexism, and classism that are wrapped up in culture and that cannot be solved but only *resolved*.[6] CTDA thus relies on and combines with critical race theory, queer theory, critical feminist studies, and the like. CTDA centralizes marginalized persons and groups, bringing underrepresented voices to the fore and attending to intersecting power structures that inform and take shape through digital products and processes.[7]

CTDA provides a strong foundation for interrogating *how* internet technologies afford across subjects and circumstances. Rather than a singular internet or depiction of a monolithic interface experience, CTDA intrinsically asks *for whom and under what circumstances?*, escaping general statements about "what people do online" and specifying the ways artifacts and ideology intersect with agentic and socially situated subjects.[8] CTDA creates a critical empirical orientation, and the mechanisms and conditions framework puts that orientation into practice.

The Walkthrough Method and App Feature Analysis

Both of the next two approaches are specific to studies of software applications (apps). Apps are software programs that serve a singular, specific purpose.[9] Although apps are most commonly associated with mobile devices, they are also incorporated into desktop and laptop interfaces. Apps represent a significant part of the contemporary digital landscape and make up an immense economy of exchange between corporate entities, developers, and users in which both money and data serve as currency.

I present the walkthrough method and app feature analysis together because of their shared empirical target. Although both analyze apps, the two approaches operate with different units of analysis and offer complementary strengths. The walkthrough method takes single apps as the unit of analysis. It entails a deep dive into the operation of a given app as that app takes shape through user publics. In contrast, the unit of analysis for app feature analysis is an entire genre of apps (such as health apps, parenting apps, privacy apps, or gambling apps). App feature analysis takes a bird's-eye view to map a broad technological and ideological landscape. In short, the walkthrough method interrogates individual apps, and app feature analysis assembles a dataset from a corpus of similarly themed apps. Apps are the data points in app feature analysis, and in the walkthrough method, each app is the data.

The Walkthrough Method

The walkthrough method is a synthesis of critical technology studies with traditional cultural studies techniques by which artifacts are read as texts.[10] The walkthrough method interrogates single apps or small groups of apps to discern how the software intersects with, reinforces, and potentially diverges from normative cultural standards in the hands of user publics. The walkthrough method involves deep engagement between the researcher and the technology of interest. Although the method can be used in conjunction with interviews and other forms of user-experience elicitation, the walkthrough method does not itself include user-experience data. The walkthrough method for app analysis builds on traditional "walkthroughs" in the field of engineering, which aim at improving design for diverse imagined users.[11] The walkthrough method discussed herein includes an explicitly political element that has been adapted for the specific study of software applications.

The walkthrough method involves three broad stages: registration and entry; everyday use; and suspension, closure, and leaving.[12] The researcher carefully documents and analyzes the technical components of registering for an app, utilizing its features, and attempting to disengage. These technical processes are situated and multiply interpreted through critical theoretical frameworks (including critical race studies, queer theory, critical STS, and critical feminism) to reveal the cultural and political underpinnings of an app's interface. The researcher approaches the app not as a static article but as a dynamic creation that will take shape in expected and unexpected ways for different users.

The goal of the walkthrough method is to identify the invisible infrastructure of an app by which technical systems quietly but effectively generate products, actions, and ideologies.[13] Unearthing an app's invisible infrastructure helps paint an image of the app's *environment of expected use*, defined as the way an "app provider anticipates [the app] will be received, generate profit or other forms of benefit, and regulate user activity."[14] The environment of expected use is broken into three factors: vision, operating model, and governance.

Vision refers to the intended purpose of the app, its target user, and the presumed context of use. From the perspective of the mechanisms and conditions framework, this roughly translates into the conditions of affordance preemptively imagined by app producers (*for whom is the app intended, and under what circumstances is its use expected?*). Operating model refers to an app's business model and revenue streams. This is the political economy of the app. Governance refers to the management and regulation of user activity through technical features and terms of service. Governance is thus how an app *requests, demands, encourages, discourages, refuses,* and *allows.*

The walkthrough method resists universal renderings of app users. The method requires imagining app features and processes from multiple user perspectives and detailing the political implications therein. For instance, researchers have studied apps from a queer* perspective to discern which forms of gender identity and sexuality are deemed legitimate and which are marginalized, ignored, or actively rejected.[15] Similar analyses could derive from intersections of race, class, gender, (dis)ability, early adopters, and older adults. Moreover, the method attends to unexpected uses, giving voice to user agencies and insubordinations. Environments of expected use are thus a base form that interacts with diverse user publics in multiple ways and toward varied ends.

Actor-network theory (ANT) currently serves as the walkthrough method's theoretical underpinning. However, as discussed in the early chapters of this book, ANT maintains political neutrality in its equivalent treatment of subjects and objects, leaving little space for critical accountability. ANT's apolitical basis is incongruous with the walkthrough method's explicitly political orientation. Indeed, the walkthrough method assumes that "technologies serve the cultural aspirations of their creators, who often accrue power by oppressing particular groups."[16] In contrast, ANT's central practitioners concede that the framework cannot be evaluated on political grounds.[17] The mechanisms and conditions framework provides an operational tool that maintains political sharpness. It is therefore an effective way to organize findings from walkthrough method studies while remaining theoretically synchronic.

App Feature Analysis

App feature analysis is a newly devised approach to the study of app genres. Introduced by Rena Bivens and Amy Adele Hasinoff, app feature analysis is aimed at "uncovering the ideologies that underlie design."[18] The method includes analysis of material features, the cultural assumptions those features embody, and a future-oriented imagining of research-based redesign. The method is less about identifying the specific features of a single application and more about revealing trends through the broad analysis of an app genre. Tracing an entire genre, rather than focusing on singular products, uncovers cultural norms as they manifest in the mundane and widespread technologies that permeate daily life.

App feature analysis is a mixed-methods approach that employs both quantitative and interpretive techniques. By documenting and interpreting technical features across multiple products, app feature analysis maps a sociotechnical landscape. Bivens and Hasinoff conducted a case study of antirape apps to demonstrate the method.[19] They identified 215 mobile apps intended to mitigate sexual violence. These 215 apps make up the authors' dataset. Within the dataset, they identified 807 features. For each feature, they documented the actions enabled, the type of violence prevention strategies, and the expected user's relationship to sexual violence (victim, perpetrator, or bystander). Their analysis yielded a powerful if disheartening finding: antirape apps reinforce victim blaming and myths of strangers as primary perpetrators of assault. The apps' orientations go against feminist projects of victim empowerment and against social science research findings that consistently show that most victims know their sexual assailants.[20]

One of the most compelling elements of app feature analysis is its movement beyond critique. Having identified trends in the application landscape, the method entails a productive reimagining. Thus, app feature analysis is not only deconstructive but reconstructive, too. For example, the authors imagine apps that would collect and distribute women's stories of assault. Apps such as these would mobilize women's narratives to create a critical threshold at which the magnitude of assault could not escape public scrutiny and would help the women who participate (by sharing or simply consuming) to find empowerment in shared experience.

App feature analysis is rooted in critical perspectives of design. The authors approach their topic—software applications—with the assumption that technology "reproduces culture and, in turn, influences users." Treating apps as sociocultural artifacts, app feature analysis uncovers "the social and political currents that are translated into technology design."[21]

Affordances are an integral part of app feature analysis. Indeed, affordances are explicit in Biven and Hasinoff's definition of a feature as "an action, option, or setting afforded by the mobile app and accessible to the user."[22] The mechanisms of affordance give voice to opportunities and constraints in both app analysis and reimagined design. For instance, we might restate the authors' findings to say that existing antirape apps *encourage* women to protect themselves from strangers while *allowing* men to proceed unaffected. Reimagined applications in which women aggregate their collective stories would instead *demand* public attention and *refuse* public ignorance toward gendered sexual violence.

The authors already begin to account for the conditions of affordance in their analytic categories of intended users (victim, perpetrator, or bystander). That is, the authors attend to the question of *for whom?*. App feature analysis would be further strengthened with analytic categories tied to the circumstances in which apps are employed. For instance, are apps intended for use during an assault encounter as a means of resistance, before the encounter as a preventative measure, or after the encounter as a form of documentation? The authors detail these issues, but the mechanisms and conditions framework would provide a systematic means by which to do so.

Although app feature analysis is relatively new and has not yet been applied widely, it has expansive potential to address a range of app categories (such as police accountability apps, antiracism apps, hook-up apps, health apps, and privacy apps). Moreover, the principles of app feature analysis may well be extended to other objects of study. For instance, researchers may study the material and social features of dating platforms, crowdfunding platforms, video streaming services, or news sites. App feature analysis thus shows promise, which can be bolstered in combination with the mechanisms and conditions framework.

Values Reflection

Values reflection is not a single method but an umbrella term for the suite of techniques aimed at incorporating values considerations into the design process.[23] Values reflection is production-facing. That is, the method entails engagement with stakeholders who make, commission, distribute, and implement technical products. However, the method maintains consideration for users because values reflection involves imagining how user publics will experience the technology and the range of social effects the technology will engender.

Methods of values reflection begin with the assumption that technologies are value-laden and political and that considerations of politics and values in the design process can mitigate harms and optimize benefits—including the capacity to define what is harmful, what is beneficial, and for whom. Values reflection is inherently imaginative. It is based on exercises that help technology producers envisage how the product will take shape across contexts. The mechanisms and conditions framework could give structure to these imaginings while maintaining space for flexibility and argumentation. Thus, engineers might imagine how their products *request* versus *demand* compliance from different potential subjects and to what ends. If effects are deleterious under some conditions, practitioners can change course.

Values reflection techniques include a range of strategies that help practitioners identify the values that inform design and imagine how these values will translate into user experiences. Some common techniques include value dams and flows, mock-ups, prototypes, field deployments, and values scenarios.[24] Dams and flows identify value tensions and work to reduce them in the design space. This includes removing design elements that even small contingents find highly objectionable (value dams) and centralizing those elements that a substantial proportion of stakeholders find especially attractive (value flows). Mock-ups, prototypes, and field deployments are small-scale renderings of in-process products that offer a preview of their implementation through engagement with producers and potential users. Values scenarios entail consideration of technical objects in practice using narrative form. This aids in the process of imagining diverse users and contexts and emphasizes long- and short-term effects of technologies as they take shape through user publics.

Affordances have been integral to values-reflection projects, with researchers urging producers to consider how the features of their design shape user practices and social dynamics.[25] Values reflection is premised on the idea that identifying multiple and competing values fosters thoughtful and intentional design. The mechanisms and conditions framework builds on the existing use of affordances in values reflection methods, adding a systematic way to articulate value implications for direct and indirect stakeholders who engage technologies under an array of circumstances.

The mechanisms and conditions framework may be especially useful for infrastructure design in which size and scope pose particular challenges.[26] Infrastructures are large-scale ventures incorporating multiple technological objects, layers of production, and broadly defined stakeholders. The mechanisms and conditions framework guides and structures the complex process of identifying persons and contexts for which an infrastructure will hold relevance. That is, the mechanisms and conditions framework provides a systematic scaffold when size and scope make it difficult to pin down clear boundaries.

Adversarial Design

Adversarial design is a form of critical design founded on the principle of agonism.[27] Critical design is a means of material making that highlights and acts on the taken-for-granted assumptions and practices that organize social life.[28] Agonism is a political principle that embraces dissensus and contestation. Rather than a converging public sphere, agonism envisions politics as an ongoing process of contradiction and disagreement.[29]

Instantiating agonism through material artifacts is the main project of adversarial design, as articulated by Carl DiSalvo in his titular text.[30] Adversarial design *does the work* of agonism, giving material form to polyvalent debate. Adversarial design is explicitly political design. It addresses the values and agendas embedded in social and technical systems, makes those dynamics visible, and asserts alternative configurations.

Adversarial design takes form through three broad tactics: revealing hegemony, reconfiguring the remainder, and articulating collectives. Revealing hegemony refers to identifying and challenging intersecting nexuses of power that organize existing political arrangements and then asking whose interests and values are currently reflected and served and how built artifacts might rework these interests and values in alternative ways. Reconfiguring the remainder gives close attention to inclusion and exclusion of built features, attending to the political implications of additions and omissions. Articulating collectives does the work of agonism by constructing human and nonhuman networks that "participate together in making, exploring, and contesting alternatives to a wide variety of societal issues and conditions."[31]

Adversarial design is both an inquiry and a practice. As inquiry, adversarial design materializes vague and nebulous situational meanings so that they can be sensed and made sense of. In practice, adversarial design takes the material forms of agonism and renders them actionable. For instance, DiSalvo references the "CCD-me-not Umbrella," a surveillance-deflecting device that obstructs charged coupled device (CCD) surveillance cameras.[32] The agonistic design of the umbrella model clarifies conditions of mass surveillance while creating a material means for resistance. Even without commercial distribution, the surveillance-defying umbrella shows an alternative to ubiquitous watching.

The political orientation of adversarial design as method coincides with the political orientation of the mechanisms and conditions as an analytic frame. Adversarial design seeks out intersecting power dynamics and rearticulates them in new form. The mechanisms and conditions framework provides a vocabulary for these rearticulations. For instance, when "reconfiguring the remainder," adversarial designers might ask how inclusions and exclusions *request* compliance, *demand* subservience, or *allow* resistance. In this way, hegemony can be revealed with clarity by attending to *for whom and under what circumstances* such *requests, demands, allowances,* and so on take form.

A second convergence between the mechanisms and conditions framework of affordances and adversarial design is their shared orientation toward process and argument. Adversarial design is built on the premise of ongoing contestation. No design project reaches an ultimate political conclusion but raises questions and critiques that remain always unresolved. The unresolved nature of contestation generates a dynamic and productive political landscape. In this way, the mechanisms and conditions framework is inherently nondetermined. The boundaries between each node in the model are loose, pliable, and up for debate. No object fits neatly within a single mechanism, and conditions are always subject to change. Thus, both adversarial design and the mechanisms and conditions framework remain projects of argumentation rather than missions of fact.

Chapter Summary

This final substantive chapter addresses affordances in practice. The mechanisms and conditions framework is an analytic tool. I demonstrate here its flexibility in combination with a range of methodological strategies and orientations. The mechanisms and conditions framework is not tied to any one discipline, empirical subject matter, or methodological practice. Rather, it extends across fields, topics, and modes of knowing. The methods discussed in this chapter are merely a sampling of potential unions between theory and praxis. Creative researchers can, and I hope will, implement the mechanisms and conditions framework through their own method of choice.

Each method discussed above adheres to a clear set of criteria, making them all appropriate vehicles for critical affordance analyses. Each method centralizes political dynamics; gives voice to marginal populations and groups; maintains a reflexive orientation; assumes multiplicity of meaning, experience, and outcome; and treats materiality as consequential but not determinative. One other crucial element these methods hold in common is their focus on imagined users. Like Norman, the approaches addressed in this chapter envisage practitioners and analysts as relational subjects whose job entails understanding the world from multiple other perspectives. This process, which sociologists refer to as "role-taking,"[33] is of particular relevance given status patterns in which those who make, sell, distribute, and evaluate technologies often hold positions of privilege and maintain disproportionate access to cultural, social, and financial capital. Left unchecked, producers are likely to make products for users who are just like themselves. Understanding the reverberations of sociotechnical systems as they affect marginalized groups requires systematic attention and concerted intention. Thus, critical methodologies and critical conceptual frameworks are invaluable for uncovering and undermining power dynamics that would otherwise reproduce in material form. Simply put, critical methods and theory are necessary to overcome the white guy problem of Silicon Valley and the ivory tower of academe.

7 Conclusion

From humble beginnings as "what things furnish, for good or ill,"[1] the concept of affordance has taken on a robust and complex intellectual life. Over the course of this book, I have traced the concept's journey and tried to give "affordance" new legs and a fresh perspective. The purpose of this book has been twofold. First, the text brings together vast, diverse, and sometimes divergent treatments of affordances across disciplines and between scholars, housing them all together in a way that clarifies rather than complicates. The second purpose of the book, the main purpose, has been to explicate the mechanisms and conditions framework. The mechanisms and conditions framework reorients the driving question of affordance analyses from *what* artifacts afford to *how* they afford, attending to variations across subjects and circumstances. It offers a simple vocabulary that spans disciplines, empirical objects of study, and various goals of both analysis and design.

Over fifty years after affordance's original formulation, this book gives the concept a much-needed makeover. Having been picked up, (over)theorized, and put to work toward versatile ends, it is worth pausing to reconsider what affordance analysis *can* do, what it *is* doing, and how it can do that work better. Now, in particular, is the time for such a project. Traditional sociotechnical problems (such as road systems, built infrastructures, and the ergonomics of chairs and tables) remain relevant, joined by ubiquitous digitization and advances in AI and machine learning, which some herald as the next paradigmatic revolution. Entirely new concepts and theories are not always the answer. Sometimes, as with affordances, it's best to level up what is already trusted, tried, and true.

The rise of social media reinvigorated the concept of affordance and put it into action as theorists and practitioners scrambled to understand the transformative effects of digitally networked sociality. The growth of computer-mediated communication studies and "new media" scholarship gave the affordance concept a renaissance, with little time for theoretical consideration of this analytic workhorse. Today, social media have more or less settled into the societal fabric, giving way to a steady, measured, and considered treatment within the academic canon. This steady treatment deserves robust and agile analytic devices. The mechanisms and conditions framework serves this purpose. The very mundanity of digital social technologies and their global embeddedness across major institutions and intimate relations render their effects profound but increasingly less blaring.

Digital connectivity is now the water in which we swim. Understanding how various systems (and companies) nudge, push, pull, and arrange requires critical attention to dynamics that would otherwise seem inevitable and unchanging. Affordance analyses unearth and articulate the ways clicks, likes, and shares translate into commercially valuable data packages and politically exploitable information; how health apps distribute (moral) responsibility for bodily maintenance while outsourcing body knowledge; how news and information can be at once abundant and at the same time deeply partial and carefully crafted; how the convenience and pleasure of an inviting screen can also exert pressure to perform, directives to consume, and severe punishment for public missteps. The mechanisms and conditions framework clarifies these complexities, politicizes their implications, and renders them visible through a simple vocabulary and a model that adapts to and assumes variation across time, subjects, and circumstances.

While theorists and practitioners continue to figure out what it means to be always on, connected, and tracked, another sociotechnical shift—artificial intelligence—has captured the public imagination. The loftiest hopes tangle with the deepest anxieties as "smart" systems enter our homes, schools, hospitals, workplaces, and government institutions. Driven by existential and practical questions about the future of humanity, funds are pouring into the hands of researchers and practitioners for the study of artificial intelligence and machine learning systems. It is near impossible to open Twitter without encountering launch announcements for new AI centers, institutes, and collaborative working groups. I am currently part of a core team at my own institution enacting a large-scale interdisciplinary project to "humanise machine intelligence,[2]" part of an AI Meetup at a neighboring university, and on several AI-related mailing lists. I also have a shared Dropbox folder in which colleagues collaborate to keep up with the emergent AI literature, which moves far faster than any human could possibly read (but maybe AI can help us with that someday soon).

AI initiatives seem increasingly compulsory for major research universities. Cambridge has the Leverhulme Centre for the Future of Intelligence (CFI); Oxford has the Oxford Artificial Intelligence Society and the Future of Humanity Institute (along with its Centre for the Governance of AI); Stanford has the Stanford Institute for Human-Centered Artificial Intelligence; Tsinghua launched the Tsinghua University Institute for Artificial Intelligence; and New York University's AI Now Institute has been a leading force in the field. These join corporate hubs, nonprofits, and think tanks, such as Google's DeepMind, OpenAI, and the Allen Institute for Artificial Intelligence.[3]

The financial and intellectual resources channeled into these endeavors portend a horizon of profound technical and social change. My flummoxed encounter with a locked shopping cart (recounted in the opening of this book) may well become an antiquated problem of the past as refrigerators, closets, robotic warehouse workers, automated drones, and driverless cars all collaborate to ensure that my shelves are adequately stocked and my domestic needs met. The nature of work will change. The nature of governance will change. The nature of care will change. In these changes are both utopic possibilities and sobering capacities for harm. AI can be potentially time saving, money saving, and lifesaving. It can breach geographic barriers and traverse dangerous territories. It can undermine human biases and create more equitable outcomes. AI can also steal, kill, and devastate.

Although sociotechnical change is inevitable—AI is coming—how these developments take shape remains an open question but one over which we, as professionals and fellow humans, have some degree of control. Early evidence suggests that unimpeded, AI technologies will go down some troubling paths. The rollouts of AI systems have been plagued by bad news: policing algorithms that target poor communities of color, job sorting programs that penalize women, home assistants that eavesdrop on private conversations, and cars that crash into pedestrians. These outcomes are neither natural nor inevitable. There is still time to change course, but the time to get involved is now, while the foundations are still being poured. To get involved—to intervene in AI in a way that optimizes opportunities and ameliorates rather than exacerbates harm—requires clear, precise, and politically sharp conceptual tools. If these tools can traverse disciplines and aid in processes of production and distribution, all the better. The mechanisms and conditions framework positions affordance analysis to be one such critical apparatus.

The mechanisms and conditions framework rests on several assumptions: affordances are continuous rather than binary; affordances vary across persons and circumstances; subjects and objects are mutually constitutive; and subjects and objects have an asymmetrical relationship. Underlying these assumptions is a political orientation (artifacts do have politics) and an instability of analysis such that analytic designations are always up for debate. These assumptions drive a simple typology made up of a simple vocabulary that fosters complex and sophisticated understandings of, and engagement with, an array of sociotechnical systems—from the mundane and tedious to those that can revolutionize existing ways of life.

I close this book by proposing a series of five big questions tied to the uncertainties of a changing sociotechnical landscape. Along with each general prompt, I suggest smaller research queries that would benefit from systematic affordance analyses. These big questions are familiar. Little about them is epiphanic. They are questions entrenched in some of the most pressing and public issues, the ones about which news programs invite panelists and university syllabi dedicate weeks of study. The questions themselves are simple in a way that belies the complexity from which they arose, the magnitude of their implications, and the intricacies of resolving them. For these questions, I offer the mechanisms and conditions framework—an upgraded model of a trusty analytic device.

These questions are not exhaustive. Rather, they are a gesture toward putting the mechanisms and conditions framework into action. Having written a theoretical book, I am now most interested in the getting down to the business of doing. The questions I propose are not a research agenda but a spark for inspiration. They are future-looking because that's a fun place to explore, but the framework can and I hope will also address the full gamut of research goals, including the traditional, the mundane, and the all-important ordinary.

Big Question 1: How Do We Identify and Equalize Digital Inequalities?

Digital inequalities are disparities in access to, skill with, and the effects of, digital products and services. Access to digital technologies is more widespread than ever. It is tempting to interpret this as a closing of "digital divides" and reduction of inequities. However, as digital divides narrow, those that persist grow deeper.[4] When institutions and infrastructures are built on presumed access, the consequences for those without access amplify exponentially. Moreover, access alone does not resolve inequality because hardware and software intersect with race, class, gender, and sexuality, with effects that mirror the raced, classed, gendered, and heteronormative priorities of existing social arrangements. The mechanism and conditions framework can help articulate *how* inequalities distribute for subjects across social locations. This would add nuance to overly general statements about access and skill that mistakenly presume more access and more skill will necessarily result in beneficial outcomes across cases.[5] Some specific questions might include: How do school curriculums afford gendered relations to technology that translate into gendered patterns in technology-based careers—that is, why aren't more women in tech, and how can we change this through institutions of education? How do image- and text-based platforms afford engagement for queer*-identified persons? How do livestreaming features afford documentation and surveillance across social class demographics?

Big Question 2: How Do Social Media Affect Sociality and Psychological Well-being?

There has been vigorous debate in the academic literature about the effects of digital social platforms on social relationships and psychological well-being. Unsurprisingly, a review of these studies shows that the effects are far from uniform but vary with the features of the platform, the subjects who use them, and the conditions of their use.[6] There is also evidence that the "problem" of social media and psychosocial outcomes is overblown.[7] The mechanisms and conditions framework is a way to tease out these variations and answer concrete questions, such as: How do dominant social media platforms afford social connection and isolation, and for whom? How do content production and consumption afford mental well-being for traditional and non-traditional users? How do default privacy settings enable and constrain personal expression for diverse subjects?

Big Question 3: How Do Information Economies Affect Political Life?

A postbroadcast media landscape means that news and information travel through multiple and diverse sources and in multiple directions, rather than in a unidirectional line from concentrated media conglomerates to consumers.[8] This redistribution of knowledge and information can be empowering[9] and at the same time can undermine journalistic rigor and standards of trustworthiness.[10] This new media landscape has particular relevance for the flows of politics in everyday life. Candidates' carefully crafted images are vulnerable to disruption, and the veracity of political information is uncertain at best. The mechanisms and conditions framework can help make sense of these shifting circumstances as citizens engage in political life. Specific questions could include: How does Twitter afford political expression within tyrannical regimes? How do various social media platforms afford the disruption of mainstream political news reporting? How do algorithmic configurations affect political debate and discourse?

Big Question 4: How Will Driverless Cars Affect Urban Infrastructures?

Driverless cars are an emergent infrastructural advancement that, once fully implemented, will upend transportation systems in profound ways.[11] The range of diverse models for driverless car implementation and the features entailed therein paint drastically different portraits of infrastructural planning in the near future.[12] For instance, individual driverless cars look quite different from shared driverless fleets, and each requires distinct arrangements of roadways, schedules, and time. In turn, driverless vehicles will have distinct implications for a range of subjects. Driverless cars offer autonomy for previously immobilized persons (including elderly adults, people with certain disabilities, children, and those without a license). At the same time, the arrangement of vehicles and their availability may create barriers to access, thus reinforcing or even enhancing patterned exclusions from public space. The mechanisms and conditions framework lets us approach a range of questions about driverless vehicles in the city, such as: How do driverless vehicles afford autonomy for older adults? How do private vehicles and public fleets afford access to public space across demographic lines? How do the features of driverless vehicle systems reconfigure divisions between urban, rural, and suburban life?

Big Question 5: How Do Medical Technologies Afford Embodied Relations to Health?

The medical field is rife with technological advancement. Pharmaceutical companies are making compounds at rapid speeds, medical tracking technologies are an integral part of formal and informal care, the work of diagnosis and treatment is getting outsourced to automated systems, gene editing has now been approved for multiple clinical trials in the United States, and a scientist in China has already modified embryos.[13] These technologies affect treatment and care but also inevitably affect one's relationship to the body. Both self-guided and physician-imposed medical tracking systems materialize distinct definitions of wellness, morality, and governance,[14] large datasets normalize and objectify bodily processes,[15] and gene therapies trouble entrenched notions of "nature." The mechanisms and conditions framework can help answer critical questions about how medical technologies reconfigure the body, for whom, and under what circumstances. For instance, we might ask: How do self-tracking devices afford body knowledge and health practices? How do embedded devices afford patient-practitioner interactions? How do biodata databases engrain or subvert normalization of body ideals? How does the automation of medical care variously afford autonomy, constriction, access, and wellness for patients in public and private markets? How does gene adaptability afford health and wellness across class lines?

Moving Forward

I set forth the above big questions and small exemplar cases to inspire designers, makers, and social science researchers to put the mechanisms and conditions framework into practice. In chapter 2, I mentioned that affordances have been at once over- and undertheorized. I hope that this book has found a middle ground. My goal was to theorize affordances just enough so that the concept remains useful in a way that we do not need to keep coming back and fleshing out the minutia. Moving forward, I want to see the mechanisms and conditions framework of affordances in action. This means evaluating existing technologies and systems, editing those systems when appropriate, and using the framework in the design process to map power, politics, and values from the onset.

Part of moving forward includes analyses of multifaceted assemblages. Although this book has focused on human-technology relations, the mechanisms and conditions framework assumes and is readily applicable to complex groupings that include technology-technology relations. Sociotechnical systems entail engagement between humans and machines and also between multiple technical elements. I focus the book around human-technology relations for reasons of simplicity. As an introduction to the mechanisms and conditions framework, the goal was to highlight *how* technologies afford in socially and politically relevant ways. To present a new orienting question and analytic framework, I selected relatively simple examples that show the relation between human subjects and technological objects.

However, I remain keenly aware that assemblages are rarely only two-part systems and that technologies intertwine with each other. For instance, the simple act of writing on paper includes relationships between a writing subject, the pen, the ink, the paper, the table on which the paper rests, and myriad other apparatuses. The pen *encourages* writing for the subject, and the paper also *requests* visibility for the pen. In turn, the table *encourages* stability for the paper, without which the paper, with its flimsy material makeup, would *discourage* transfer from both pen and ink. The point is that technologies are multiply relational. This is not a new point. Multifaceted assemblages are integral in science and technology studies more generally and within affordance theory, in particular. The mechanisms and conditions framework can elucidate multifaceted assemblages in the same ways I've demonstrated throughout this text with primarily dyadic examples of human-technology relations.

The mechanisms and conditions framework attends to the complexities of sociotechnical systems, in their various forms. It does so with power and politics at the center. The framework moves along with sociotechnical changes and sets a shared vocabulary for argumentation. With the mechanisms and conditions framework, analysts and practitioners can vigorously debate about the implications of sociotechnical change and the appropriate pathways forward. The mechanisms and conditions framework is a simple tool that packs a big punch. Having laid out the model, it is now time to get to work.

Notes

Acknowledgments

1. Jenny Davis, "Theorizing Affordances," *Cyborgology* (blog), *The Society Pages*, February 16, 2015, https://thesocietypages.org/cyborgology/2015/02/16/theorizing-affordances; Jenny Davis, "A Short History of Affordances," *Cyborgology* (blog), *The Society Pages*, July 13, 2015, https://thesocietypages.org/cyborgology/2015/07/13/a-short-history-of-affordances.

2. Jenny L. Davis and James B. Chouinard, "Theorizing Affordances: From Request to Refuse," *Bulletin of Science, Technology & Society*, 36, no. 4 (June 16, 2017): 241–248, https://journals.sagepub.com/doi/full/10.1177/0270467617714944.

3. The Autonomy, Agency and Assurance Innovation (3Ai) Institute, Australian National University, Canberra, https://3ainstitute.cecs.anu.edu.au.

4. Noah Jerome Springer, "Publics and Counterpublics on the Front Page of the Internet: The Cultural Practices, Technological Affordances, Hybrid Economics and Politics of Reddit's Public Sphere" (PhD diss., Department of Journalism and Mass Communication, University of Colorado, 2015), https://search.proquest.com/docview/1719155030?pq-origsite=gscholar.

5. *Cyborgology* (blog), *The Society Pages*, https://thesocietypages.org/cyborgology.

6. Theorizing the Web (conference), https://theorizingtheweb.org.

Chapter 1

1. Canberra is relatively small for a capital city and has a suburban feel. When I moved there in 2017, coin-locked shopping carts were still new. Although they felt unusual to me, my husband, who grew up in New York City, used them regularly in the stores where his family shopped. He was amused by my confusion and slow uptake.

2. For literature on the merging of productive labor with consumption practices ("prosumption"), see Alvin Toffler, *The Third Wave* (New York: Bantam Books, 1980); George Ritzer, Paul Dean, and Nathan Jurgenson, "The Coming of Age of the Prosumer," *American Behavioral Scientist* 56, no. 4 (2012): 379–398; George Ritzer and Nathan Jurgenson, "Production, Consumption, Prosumption: The Nature of Capitalism in the Age of the Digital 'Prosumer,'" *Journal of Consumer Culture* 10, no. 1 (2010): 13–36.

3. This history of the shopping cart was constructed using historical analysis and patent records from the following sources: Jacques Ricouard and Claude Chappoux, "Coin Lock Device for Shopping Trolleys," *Official Gazette of the United States Patent and Trademark Office*, vol. 1074, issue 3, January 20, 1987, p. 1449 (Google Patents, 1987); Aage Lenander, "Coin-Operated Lock for a Trolley System Including Especially Shopping and Luggage Trolleys," *Official Gazette of the United States Patent and Trademark Office*, vol. 1047, issue 1, October 2, 1984, p. 142 (Google Patents, 1984); Frederik R. L. Rheeder and Deon Dixon, "Trolley Locking Device," *Official Gazette of the United States Patent and Trademark Office*, vol. 1064, issue 1, March 4, 1986, p. 133 (Google Patents, 1986); Ellen Ruppel Shell, *Cheap: The High Cost of Discount Culture* (New York: Penguin, 2009); Sylvan N Goldman, "Commodity Accommodation and Vending Rack," *Official Gazette of the United States Patent and Trademark Office*, vol. 1074, issue 3, January 20, 1987, p. 1449 (Google Patents,

1938); Franck Cochoy, "Driving a Shopping Cart from STS to Business, and the Other Way Round: On the Introduction of Shopping Carts in American Grocery Stores (1936–1959)," *Organization* 16, no. 1 (2009): 31–55.

4. Samer Faraj and Bijan Azad, "The Materiality of Technology: An Affordance Perspective," *Materiality and Organizing: Social Interaction in a Technological World*, ed. Paul M. Leonardi, Bonnie A. Nardi, and Jannis Kallinikos (Oxford: Oxford University Press, 2012), 254

5. Sandra K. Evans, Katy E. Pearce, Jessica Vitak, and Jeffrey W. Treem, "Explicating Affordances: A Conceptual Framework for Understanding Affordances in Communication Research," *Journal of Computer-Mediated Communication* 22, no. 1 (2017): 36.

6. James J. Gibson, *The Senses Considered as Perceptual Systems* (Oxford: Houghton Mifflin, 1966); James J. Gibson, *The Ecological Approach to Visual Perception: Classic Edition* (New York: Psychology Press, 2014).

7. Donald A. Norman, *The Psychology of Everyday Things* (New York: Basic Books, 1988); Donald A. Norman, *The Design of Everyday Things* (Cambridge, MA: MIT Press, 1998).

8. Donald A. Norman, "The Way I See It: Signifiers, Not Affordances," *Interactions* 15, no. 6 (2008): 18–19; Martin Oliver, "The Problem with Affordance," *E-Learning and Digital Media* 2, no. 4 (2005): 402–413.

9. The first articulation of the mechanisms and conditions framework can be found in Jenny L. Davis and James B. Chouinard, "Theorizing Affordances: From Request to Refuse," *Bulletin of Science, Technology & Society* 36, no. 4 (2016): 241–248.

10. For this Carnegie Mellon University study, see Amit Datta, Michael Carl Tschantz, and Anupam Datta, "Automated Experiments on Ad Privacy Settings," *Proceedings on Privacy Enhancing Technologies* 2015, no. 1 (2015): 92–112.

11. Virginia Eubanks, *Automating Inequality: How High-Tech Tools Profile, Police, and Punish the Poor* (New York: St. Martin's Press, 2018).

12. Safiya Umoja Noble, *Algorithms of Oppression: How Search Engines Reinforce Racism* (New York: NYU Press, 2018).

13. Batya Friedman, "Value-Sensitive Design," *Interactions* 3, no. 6 (1996): 16–23; Batya Friedman, P. Kahn, and Alan Borning, "Value Sensitive Design and Information Systems," in *Human-Computer Interaction In Management Information Systems: Foundations*, ed. Ping Zhang and Dennis F. Galletta, 348–372 (New York: Routledge, 2006); Batya Friedman, Peter H. Kahn, Alan Borning, and Alina Huldtgren, "Value Sensitive Design and Information Systems," in *Early Engagement and New Technologies: Opening Up the Laboratory*, ed. Neelke Doorn, Dean Schuurbiers, Ibo van de Poel, and Michael E. Gorman, 55–95 (Dordrecht: Springer, 2013); Jeroen Van der Hoven and Noemi Manders-Huits, *Value-Sensitive Design* (Hoboken, NJ: Wiley, 2009); Till Winkler and Sarah Spiekermann, "Twenty Years of Value Sensitive Design: A Review of Methodological Practices in VSD Projects," *Ethics and Information Technology* (2018): 1–5; Batya Friedman and David G. Hendry, *Value Sensitive Design: Shaping Technology with Moral Imagination* (Cambridge, MA: MIT Press, 2019). See Value Sensitive Design Research Lab, Information School and Department of Computer Science and Engineering, University of Washington, https://vsdesign.org, for an overview and further relevant works.

14. Mary Flanagan and Helen Nissenbaum, *Values at Play in Digital Games* (Cambridge, MA: MIT Press, 2014).

15. Eubanks, *Automating Inequality*, 11.

16. The political economy of technology is a robust and longstanding field in the social sciences. These two works were chosen for their exemplary quality and contemporary relevance. For more examples from this tradition see: Siva Vaidhyanathan, *Antisocial Media: How Facebook Disconnects Us and Undermines Democracy* (New York: Oxford University Press, 2018); Cathy O'Neil, *Weapons of Math Destruction: How Big Data Increases Inequality and Threatens Democracy* (New York: Broadway Books, 2016); Paul Dourish, *The Stuff of Bits: An Essay on the Materialities of Information* (Cambridge, MA: MIT Press, 2017); Langdon Winner, "Do Artifacts Have Politics?," *Daedalus* 109, no. 1 (1980): 121–136; Steve

Woolgar and Geoff Cooper, "Do Artefacts Have Ambivalence? Moses' Bridges, Winner's Bridges and Other Urban Legends in S&Ts," *Social Studies of Science* 29, no. 3 (1999): 433–449; Lucas Introna and David Wood, "Picturing Algorithmic Surveillance: The Politics of Facial Recognition Systems," *Surveillance & Society* 2, no. 2/3 (2002); Kate Crawford, "Can an Algorithm Be Agonistic? Ten Scenes from Life in Calculated Publics," *Science, Technology & Human Values* 41, no. 1 (2016): 77–92; Frank Pasquale, *The Black Box Society: The Secret Algorithms That Control Money and Information* (Cambridge, MA: Harvard University Press, 2015); Marc Berg, "The Politics of Technology: On Bringing Social Theory into Technological Design," *Science, Technology & Human Values* 23, no. 4 (1998): 456–490; Judy Wajcman, "The Gender Politics of Technology," in *The Oxford Handbook of Contextual Political Analysis*, ed. Robert E. Goodin and Charles Tilly, 707–721 (Oxford: Oxford University Press, 2006); Ruha Benjamin, "Catching Our Breath: Critical Race STS and the Carceral Imagination," *Engaging Science, Technology, and Society* 2 (2016): 145–156.

17. Noble, *Algorithms of Oppression*, 90.

18. Winkler and Spiekermann, "Twenty Years of Value Sensitive Design."

19. Bruno Latour, *Reassembling the Social: An Introduction to Actor-Network-Theory* (New York: Oxford University Press, 2005)

20. Ernst Schraube, "Technology as Materialized Action and Its Ambivalences," *Theory & Psychology* 19, no. 2 (2009): 296–312.

Chapter 2

1. For critiques of the misuse, overuse, and undertheorization of affordances, see Leonardo Burlamaqui and Andy Dong, "The Use and Misuse of the Concept of Affordance," in *Design Computing and Cognition* , ed. John S. Gero and Sean Hanna, 295–311 (New York: Springer, 2015); Sandra K. Evans, Katy E. Pearce, Jessica Vitak, and Jeffrey W. Treem, "Explicating Affordances: A Conceptual Framework for Understanding Affordances in Communication Research," *Journal of Computer-Mediated*

Communication 22, no. 1 (2017): 35–52; Tim Ingold, "Back to the Future with the Theory of Affordances," *HAU: Journal of Ethnographic Theory* 8, no. 1–2 (2018): 39–44; Keith S. Jones, "What Is an Affordance?," *Ecological Psychology* 15, no. 2 (2003):107–114; Joanna McGrenere and Wayne Ho, "Affordances: Clarifying and Evolving a Concept," *Proceedings of the Graphics Interface 2000: Montreal, Quebec, Canada, 15–17 May 2000* (Montreal: Canadian Human-Computer Communications Society, 2000); Martin Oliver, "The Problem with Affordance," *E-Learning and Digital Media* 2, no. 4 (2005): 402–413; Donald A. Norman, "The Way I See It: Signifiers, Not Affordances," *Interactions* 15, no. 6 (2008): 18–19.

2. James J. Gibson, *The Senses Considered as Perceptual Systems* (Boston: Houghton Mifflin, 1966); James J. Gibson, *The Ecological Approach to Visual Perception* (Boston: Houghton Mifflin 1979).

3. Russell Kahl, ed., *Selected Writings of Hermann Von Helmholtz* (Middletown, CT: Wesleyan University Press, 1971).

4. James J. Gibson, *The Ecological Approach to Visual Perception: Classic Edition* (New York: Psychology Press, 2014).

5. Edward S. Reed, "The Affordances of the Animate Environment: Social Science from the Ecological Point of View," in *What Is an Animal?*, ed. Tim Ingold, 110–126 (Abingdon, UK: Routledge, 1988).

6. Gibson, *The Senses Considered as Perceptual Systems*, 285

7. Gibson, *The Ecological Approach to Visual Perception*, 127 (emphasis in the original).

8. Dobromir G. Dotov, Lin Nie, and Matthieu M. De Wit, "Understanding Affordances: History and Contemporary Development of Gibson's Central Concept," *Avant: The Journal of the Philosophical-Interdisciplinary Vanguard* 3, no. 2 (2012): 30.

9. Gibson, *The Ecological Approach to Visual Perception*, 127.

10. Kurt Koffka, *Principles of Gestalt Psychology* (New York: Harcourt, 1935); Kurt Lewin, *A Dynamic Theory of Personality: Selected Papers*, trans. by Donald K. Adams and Karl E. Zener (New York: McGraw, 1935).

11. Gibson, *The Ecological Approach to Visual Perception*, 130.

12. Gibson, *The Ecological Approach to Visual Perception*, 130.

13. Donald A. Norman, *The Psychology of Everyday Things* (New York: Basic Books, 1988).

14. Norman, *The Psychology of Everyday Things*, 9

15. Donald A. Norman, *The Design of Everyday Things* (Cambridge, MA: MIT Press, 1998).

16. See Oliver, "The Problem with Affordance."

17. Norman, *The Design of Everyday Things*, 41.

18. Ingold, "Back to the Future with the Theory of Affordances"; Tim Ingold, *Being Alive: Essays on Movement, Knowledge and Description* (New York: Routledge, 2011); Tim Ingold, "Culture and the Perception of the Environment," in *Bush Base, Forest Farm: Culture, Environment, and Development*, ed. Elisabeth Croll and David Parkin, 51–68 (London: Routledge, 2002); Bryan Pfaffenberger, "Social Anthropology of Technology," *Annual Review of Anthropology* 21, no. 1 (1992): 491–516.

19. Ingold, "Back to the Future with the Theory of Affordances," 40.

20. Pfaffenberger, "Social Anthropology of Technology," 497.

21. Jonathan R. A. Maier and Georges M. Fadel, "Affordance Based Design: A Relational Theory for Design," *Research in Engineering Design* 20, no. 1 (2009): 13–27; Jonathan R. A. Maier and Georges M. Fadel, "Affordance-Based Design Methods for Innovative Design, Redesign and Reverse Engineering," *Research In Engineering Design* 20, no. 4 (2009): 225; Jonathan R. A. Maier and Georges M. Fadel, "Affordance: The Fundamental Concept in Engineering Design," Paper No. Detc2001/Dtm-21700, ASME Design Theory and Methodology Conference, Pittsburgh, Pennsylvania, 2001.

22. See Samer Faraj and Bijan Azad, "The Materiality of Technology: An Affordance Perspective," in *Materiality and Organizing: Social Interaction in a Technological World*, ed. Paul M. Leonardi, Bonnie A. Nardi, and Jannis

Kallinikos, 237–258 (Oxford: Oxford University Press, 2012); William
W. Gaver, "Situating Action II: Affordances for Interaction. The Social
Is Material for Design," *Ecological Psychology* 8, no. 2 (1996): 111–129;
Tarleton Gillespie, "The Politics of 'Platforms,'" *New Media & Society* 12,
no. 3 (2010): 347–364; Ian Hutchby, "Technologies, Texts and Affordances," *Sociology* 35, no. 2 (2001): 441–456; Paul M. Leonardi, "When
Flexible Routines Meet Flexible Technologies: Affordance, Constraint,
and the Imbrication of Human and Material Agencies," *MIS Quarterly*
35, no.1 (2011): 147–167; Paul M. Leonardi, "Theoretical Foundations
for the Study of Sociomateriality," *Information and Organization* 23, no.
2 (2013): 59–76; Jeffrey W. Treem and Paul M. Leonardi, "Social Media
Use in Organizations: Exploring the Affordances of Visibility, Editability, Persistence, and Association," *Annals of the International Communication Association* 36, no. 1 (2013): 143–189.

23. See danah boyd, "Social Network Sites as Networked Publics:
Affordances, Dynamics, and Implications," in *A Networked Self*, ed.
Zizi Papacharissi, 47–66 (New York: Routledge, 2010); Andrew Richard
Schrock, "Communicative Affordances of Mobile Media: Portability,
Availability, Locatability, and Multimediality," *International Journal
of Communication* 9 (2015): 1229–1246; Jenny Davis, "Architecture
of the Personal Interactive Homepage: Constructing the Self through
Myspace," *New Media & Society* 12, no. 7 (2010): 1103–1109; Kate Sarah
Raynes-Goldie, "Privacy in the Age of Facebook: Discourse, Architecture,
Consequences" (PhD diss., Curtin University, 2012).

24. See Jennifer L. Gibbs, Nik Ahmad Rozaidi, and Julia Eisenberg,
"Overcoming the 'Ideology of Openness': Probing the Affordances
of Social Media for Organizational Knowledge Sharing," *Journal of
Computer-Mediated Communication* 19, no. 1 (2013): 102–120; Ann Maj-
chrzak, Samer Faraj, Gerald C. Kane, and Bijan Azad, "The Contradictory
Influence of Social Media Affordances on Online Communal Knowledge
Sharing," *Journal of Computer-Mediated Communication* 19, no. 1 (2013):
38–55.

25. See Kate Crawford and Tarleton Gillespie, "What Is a Flag For?
Social Media Reporting Tools and the Vocabulary of Complaint," *New*

Media & Society 18, no. 3 (2016): 410–428; Tarleton Gillespie, *Custodians of the Internet: Platforms, Content Moderation, and the Hidden Decisions That Shape Social Media* (New Haven, CT: Yale University Press, 2018); Adrienne Massanari, "# Gamergate and the Fappening: How Reddit's Algorithm, Governance, and Culture Support Toxic Technocultures," *New Media & Society* 19, no. 3 (2017): 329–346.

26. See André Brock, "From the Blackhand Side: Twitter as a Cultural Conversation," *Journal of Broadcasting & Electronic Media* 56, no. 4 (2012): 529–549; Jenny L. Davis, "Triangulating the Self: Identity Processes in a Connected Era," *Symbolic Interaction* 37, no. 4 (2014): 500–523; Alice E. Marwick and danah boyd, "I Tweet Honestly, I Tweet Passionately: Twitter Users, Context Collapse, and the Imagined Audience," *New Media & Society* 13, no. 1 (2011): 114–133.

27. See Gale Parchoma, "The Contested Ontology of Affordances: Implications for Researching Technological Affordances for Collaborative Knowledge Production," *Computers in Human Behavior* 37, no. Supp. C (2014): 360–368; Steve Wright and Gale Parchoma, "Technologies for Learning? An Actor-Network Theory Critique of 'Affordances' in Research on Mobile Learning," *Research in Learning Technology* 19, no. 3 (2011): 247–258, http://dx.doi.org/10.1080/21567069.2011.624168.

28. Roy D. Pea, "Practices of Distributed Intelligence and Designs for Education," *Distributed Cognitions: Psychological and Educational Considerations* 11 (1993): 47–87.

29. Diana Laurillard, Matthew Stratfold, Rose Luckin, Lydia Plowman, and Josie Taylor, "Affordances for Learning in a Non-Linear Narrative Medium," *Journal of Interactive Media in Education*, no. 2 (2000): 1–19.

30. Daniel D. Suthers, "Technology Affordances for Intersubjective Meaning Making: A Research Agenda for CSCL," *International Journal of Computer-Supported Collaborative Learning* 1, no. 3 (2006): 315–337.

31. Grainne Conole and Martin Dyke, "What Are the Affordances of Information and Communication Technologies?," *ALT-J Research in Learning Technology* 12, no. 2 (2004): 113–124.

32. William M. Mace, "James J. Gibson's Ecological Approach: Perceiving What Exists," *Ethics and the Environment* 10, no. 2 (2005): 195–216.

33. William H. Warren, "Perceiving Affordances: Visual Guidance of Stair Climbing," *Journal of Experimental Psychology: Human Perception and Performance* 10, no. 5 (1984): 683–703.

34. See Claire F. Michaels, "Affordances: Four Points of Debate," *Ecological Psychology* 15, no. 2 (2003): 135–148; Claire F. Michaels and Claudia Carello, *Direct Perception* (Englewood Cliffs, NJ: Prentice-Hall, 1981); Robert Shaw, "Ecological Psychology: The Consequence of a Commitment to Realism," *Cognition and the Symbolic Processes*, ed. Walter B. Weimer and David S. Palermo, 159–226 (Hillsdale, NJ: Lawrence Erlbaum, 1982); Michael T. Turvey, "Affordances and Prospective Control: An Outline of the Ontology," *Ecological Psychology* 4, no. 3 (1992): 173–187.

35. Dotov, Nie, and De Wit, "Understanding Affordances,"31.

36. See James E. Cutting, "Two Ecological Perspectives: Gibson vs. Shaw and Turvey," *American Journal of Psychology* 95, no. 2 (1982): 199–222.

37. Turvey, "Affordances and Prospective Control," 180.

38. McGrenere and Ho, "Affordances."

39. Benjamin T. Ciavola and John K. Gershenson, "Affordance Theory for Engineering Design," *Research in Engineering Design* 27, no. 3 (2016): 251–263.

40. Ciavola and Gershenson, "Affordance Theory for Engineering Design," 254.

41. Anthony Chemero, "An Outline of a Theory of Affordances," *Ecological Psychology* 15, no. 2 (2003): 181–195.

42. Hutchby, "Technologies, Texts and Affordances," 415.

43. Andrea Scarantino, "Affordances Explained," *Philosophy of Science* 70, no. 5 (2003): 960.

44. Richard C. Schmidt, "Scaffolds for Social Meaning," *Ecological Psychology* 19, no. 2 (2007):137.

45. Lee Humphreys, Veronika Karnowski, and Thilo von Pape, "Smartphones as Metamedia: A Framework for Identifying the Niches Structuring Smartphone Use," *International Journal of Communication* 12 (2018): 2793–2809); Maier and Fadel, "Affordance Based Design"; Parchoma, "The Contested Ontology of Affordances"; Dhaval Vyas, Cristina M. Chisalita, and Alan Dix, "Organizational Affordances: A Structuration Theory Approach to Affordances," *Interacting with Computers* 29, no. 2 (2017): 117–131.

46. Leslie Z. McArthur and Reuben M. Baron, "Toward an Ecological Theory of Social Perception," *Psychological Review* 90, no. 3 (1983): 215.

47. Schmidt, "Scaffolds for Social Meaning."

48. Leonardi, "When Flexible Routines Meet Flexible Technologies"; Leonardi, "Theoretical Foundations for the Study of Sociomateriality"; Treem and Leonardi, "Social Media Use in Organizations."

49. See Jenny L. Davis and James B. Chouinard, "Theorizing Affordances: From Request to Refuse," *Bulletin of Science, Technology & Society* 36 no. 4 (2016): 241–248; Evans, Pearce, Vitak, and Treem, "Explicating Affordances"; Jones, "What Is an Affordance?"; McGrenere and Ho, "Affordances"; Oliver, "The Problem with Affordance"; Thomas A. Stoffregen, "Affordances as Properties of the Animal-Environment System," *Ecological Psychology* 15, no. 2 (2003): 115–134; Gerard Torenvliet, "We Can't Afford It! The Devaluation of a Usability Term," *Interactions* 10, no. 4 (2003): 12–17.

50. Ingold, "Back to the Future with the Theory of Affordances," 39.

51. Norman, "The Way I See It: Signifiers, Not Affordances."

52. See Evans, Pearce, Vitak, and Treem, "Explicating Affordances"; Hutchby, "Technologies, Texts and Affordances"; McGrenere and Ho, "Affordances"; Peter Nagy and Gina Neff, "Imagined Affordance: Reconstructing a Keyword for Communication Theory," *Social Media + Society* 1, no. 2 (2015); Gina Neff, Tim Jordan, Joshua McVeigh-Schultz, and

Tarleton Gillespie, "Affordances, Technical Agency, and the Politics of Technologies of Cultural Production," *Journal of Broadcasting & Electronic Media* 56, no. 2 (2012): 299–313.

53. McGrenere and Ho, "Affordances," 7.

54. See Chemero, "An Outline of a Theory of Affordances"; McGrenere and Ho, "Affordances"; Nagy and Neff, "Imagined Affordance"; Neff, Jordan, McVeigh-Schultz, and Gillespie, "Affordances, Technical Agency, and the Politics of Technologies of Cultural Production"; Scarantino, "Affordances Explained."

55. Nagy and Neff, "Imagined Affordance."

56. Evans, Pearce, Vitak, and Treem, "Explicating Affordances."

57. Jun Hu and George M. Fadel, "Categorizing Affordances for Product Design" (paper presented at the ASME 2012 International Design Engineering Technical Conferences and Computers and Information in Engineering Conference, August 2012); Maier and Fadel, "Affordance-Based Design Methods for Innovative Design, Redesign and Reverse Engineering"; Ivan Mata, Georges Fadel, and Gregory Mocko, "Toward Automating Affordance-Based Design," *AI EDAM: Artificial Intelligence for Engineering Design, Analysis and Manufacturing* 29, no. 3 (2015): 297–305.

58. Davis and Chouinard, "Theorizing Affordances: From Request to Refuse."

59. Brooke Dinsmore, "Contested Affordances: Teachers and Students Negotiating the Classroom Integration of Mobile Technology," *Information, Communication & Society* 22, no. 5 (2019): 664–677; Kate Mannell, "A Typology of Mobile Messaging's Disconnective Affordances," *Mobile Media & Communication* 7, no. 1 (2019): 76–93; Apryl A. Williams, Zaida Bryant, and Christopher Carvell, "Uncompensated Emotional Labor, Racial Battle Fatigue, and (In)Civility in Digital Spaces," *Sociology Compass* 13, no. 2 (2019): e12658.

60. Jennifer C. Mueller, "Racial Ideology or Racial Ignorance? An Alternative Theory of Racial Cognition," Open Science Framework, March 2, 2019, https://osf.io/fw23k.

Chapter 3

1. Taina Bucher, *If . . . Then: Algorithmic Power and Politics* (New York: Oxford University Press, 2018).

2. See Lucas D. Introna, "The Enframing of Code: Agency, Originality and the Plagiarist," *Theory, Culture & Society* 28, no. 6 (2011): 113–141; Bruno Latour, *Reassembling the Social: An Introduction to Actor-Network Theory* (New York: Oxford University Press, 2005); Martin Müller, "Assemblages and Actor-Networks: Rethinking Socio-Material Power, Politics and Space," *Geography Compass* 9, no. 1 (2015): 27–41; Phillip Vannini, "Non-Representational Research Methodologies: An Introduction," in *Non-Representational Methodologies: Reenvisioning Research*, ed. Phillip Vannini, 11–28 (New York: Routledge, 2015); Peter-Paul Verbeek, "Materializing Morality: Design Ethics and Technological Mediation," *Science, Technology & Human Values* 31, no. 3 (2006): 361–380.

3. Ernst Schraube, "Technology as Materialized Action and Its Ambivalences," *Theory & Psychology* 19, no. 2 (2009): 296–312.

4. Marshall McLuhan, *Understanding Media: The Extensions of Man* (New York: McGraw-Hill, 1964).

5. Latour, *Reassembling the Social*.

6. Schraube, "Technology as Materialized Action and Its Ambivalences."

7. Langdon Winner, "Do Artifacts Have Politics?," *Daedalus* 109, no. 1 (1980): 121–136.

8. McLuhan, *Understanding Media*.

9. See Elihu Katz and Paul F. Lazarsfeld, *Personal Influence: The Part Played by People in the Flow of Mass Communications* (New York: Free Press, 1966); Jefferson Pooley and Elihu Katz, "Further Notes on Why American Sociology Abandoned Mass Communication Research," *Journal of Communication* 58, no. 4 (2008): 767–786.

10. McLuhan, *Understanding Media*, 10.

11. See Latour, *Reassembling the Social*; Michel Callon, "The Sociology of An Actor-Network: The Case of the Electric Vehicle," in *Mapping the Dynamics of Science and Technology: Sociology of Science in the Real World*, ed. Michel Callon, John Law, and Arie Rip, 19–34 (London: Palgrave Macmillan, 1986); Michel Callon and John Law, "After the Individual in Society: Lessons on Collectivity from Science, Technology and Society," *Canadian Journal of Sociology / Cahiers Canadiens de Sociologie* 22, no. 2 (1997): 165–182; John Law, "Actor Network Theory and Material Semiotics," in *The New Blackwell Companion to Social Theory*, ed. Brian S. Turner, 141–158 (Oxford: Wiley-Blackwell, 2009).

12. Gilles Deleuze and Felix Guattari, *A Thousand Plateaus*, trans. Brian Massumi (London: Continuum, 1987).

13. David Banks, "A Brief Summary of Actor Network Theory," *Cyborgology* (blog), *The Society Pages*, 2011, https://thesocietypages.org/cyborgology/2011/12/02/a-brief-summary-of-actor-network-theory.

14. See Warwick Anderson, "From Subjugated Knowledge to Conjugated Subjects: Science and Globalisation, or Postcolonial Studies of Science?," *Postcolonial Studies* 12, no. 4 (2009): 389–400; David Bloor, "Anti-Latour," *Studies in History and Philosophy of Science Part A* 30, no. 1 (1999): 81–112; Sandra Harding, I. Grewal, C. Kaplan, and R. Wiegman, *Sciences from Below: Feminisms, Postcolonialities, and Modernities* (Durham, NC: Duke University Press, 2008); Sal Restivo, "Review Essays: Politics of Latour," *Organization & Environment* 18, no. 1 (2005): 111–115.

15. See Kim Fortun, "From Latour to Late Industrialism," *Journal of Ethnographic Theory* 4, no. 1 (2014): 309–329; Harding, Grewal, Kaplan, and Wiegman, *Sciences from Below*; Judy Wajcman, "Reflections on Gender and Technology Studies: In What State Is the Art?," *Social Studies of Science* 30, no. 3 (2000): 447–464.

16. Mark Andrejevic, "Data Collection without Limits: Automated Policing and the Politics of Framelessness," in *Big Data, Crime and Social Control*, ed. Ales Zavrsnik, 111–125 (London: Routledge, 2017); Sarah Brayne, "Big Data Surveillance: The Case of Policing," *American Sociological Review* 82, no. 5 (2017): 977–1008.

17. Richard Dyer, *White: Essays on Race and Culture*, 2nd ed. (London: Routledge, 2017).

18. Kevin Granville, "Facebook and Cambridge Analytica: What You Need to Know as Fallout Widens," *New York Times*, March 19, 2018; Siva Vaidhyanathan, *Antisocial Media: How Facebook Disconnects Us and Undermines Democracy* (Oxford: Oxford University Press, 2018).

19. Edwin Sayes, "Actor-Network Theory and Methodology: Just What Does It Mean to Say That Nonhumans Have Agency?," *Social Studies of Science* 44, no. 1 (2014): 134–149.

20. Winner, "Do Artifacts Have Politics?"

21. Selena Savić and Gordan Savičić, "Unpleasant Design: Designing Out Unwanted Behaviour," in *A Matter of Design: Making Society through Science and Technology. Proceedings of the Fifth STS Italia Conference* (Rome: Società Italiana di Studi sulla Scienza la Tecnologia, 2014).

22. Patrick Marshall, "Algorithms Can Mask Biases in Hiring," *Sage Business Researcher*, February 15, 2016.

23. Matthew Adam Bruckner, "The Promise and Perils of Algorithmic Lenders' Use of Big Data," *Chicago-Kent Law Review* 93, no. 1 (2018): 3.

24. Schraube, "Technology as Materialized Action and Its Ambivalences," 298.

25. Schraube, "Technology as Materialized Action and Its Ambivalences," 299–300.

26. Schraube, "Technology as Materialized Action and Its Ambivalences," 300.

27. Schraube, "Technology as Materialized Action and Its Ambivalences," 205.

28. Kate Crawford and Meredith Whittaker, "The AI Now Report: The Social and Economic Implications of Artificial Intelligence Technologies in the Near-Term," AI Now Institute, September 22, 2016, https://ainowinstitute.org/AI_Now_2016_Report.pdf.

29. Meredith Broussard, *Artificial Unintelligence: How Computers Misunderstand the World* (Cambridge, MA: MIT Press, 2018).

30. Frank Pasquale, "Algorithms Can Be a Digital Star Chamber," *Aeon*, no. 8 (2015), https://acon.co.essays/judge-jury-and-executioner-the -unaccountable-algorithm.

31. Verbeek, "Materializing Morality."

32. Schraube, "Technology as Materialized Action and Its Ambivalences," 306.

33. Don Ihde, *Technology and the Lifeworld: From Garden to Earth* (Bloomington: Indiana University Press, 1990).

34. Zeynep Tufekci, *Twitter and Tear Gas: The Power and Fragility of Networked Protest* (New Haven, CT: Yale University Press, 2017).

Chapter 4

1. William H. Warren, "Perceiving Affordances: Visual Guidance of Stair Climbing," *Journal of Experimental Psychology: Human Perception and Performance* 10, no. 5 (1984): 683–703.

2. Sandra K. Evans, Katy E. Pearce, Jessica Vitak, and Jeffrey W. Treem, "Explicating Affordances: A Conceptual Framework for Understanding Affordances in Communication Research," *Journal of Computer-Mediated Communication* 22, no. 1 (2017): 35–52.

3. Rob Withagen, Duarte Araújo, and Harjo J. de Poel, "Inviting Affordances and Agency," *New Ideas in Psychology* 45 (2017): 11–18.

4. For further explication of this critique, see Joanna McGrenere and Wayne Ho, "Affordances: Clarifying and Evolving a Concept," in *Proceedings of Graphics Interface 2000: Montreal, Quebec, Canada, 15–17 May 2000* (Montreal: Canadian Human-Computer Communications Society, 2000); Peter Nagy and Gina Neff, "Imagined Affordance: Reconstructing a Keyword for Communication Theory," *Social Media + Society* 1, no. 2 (2015); Gina Neff, Tim Jordan, Joshua McVeigh-Schultz, and Tarleton Gillespie, "Affordances, Technical Agency, and the Politics of

Technologies of Cultural Production," *Journal of Broadcasting & Electronic Media* 56, no. 2 (2012): 299–313; Gale Parchoma, "The Contested Ontology of Affordances: Implications for Researching Technological Affordances for Collaborative Knowledge Production," *Computers in Human Behavior* 37, no. Suppl. C (2014): 360–368; Steve Wright and Gale Parchoma, "Technologies for Learning? An Actor-Network Theory Critique of 'Affordances' in Research on Mobile Learning," *Research in Learning Technology* 19, no. 3 (2011): 247–258, http://dx.doi.org/10.108 0/21567069.2011.624168.

5. Danny Wicentowski, "Defying Police 'Do Not Cross' Line Would Be a Crime under Proposed Missouri Bill," *River Front Times* (blog), February 23, 2016, https://www.riverfronttimes.com/newsblog/2016/12/23/ defying-police-do-not-cross-line-would-be-a-crime-under-proposed-missouri-bill.

6. HB37 died in committee, leaving the offense punishable by an up to $500 fine and community service.

7. See Jefferson D. Pooley, "Open Media Scholarship: The Case for Open Access in Media Studies," *International Journal of Communication* 10 (2016): 6148–6164.

8. Rena Bivens, "The Gender Binary Will Not Be Deprogrammed: Ten Years of Coding Gender on Facebook," *New Media & Society* 19, no. 6 (2017): 880–898.

9. Ernst Schraube, "Technology as Materialized Action and Its Ambivalences," *Theory & Psychology* 19, no. 2 (2009): 296–312.

10. Brian Wansink, Koert van Ittersum, and James E. Painter, "Ice Cream Illusions: Bowls, Spoons, and Self-Served Portion Sizes," *American Journal of Preventive Medicine* 31, no. 3 (2006):240–243.

11. Portions Master, "Products: 125lbs/57kg Skinny Plate," https:// portionsmaster.com/products/view/125lbs-57kg-portions-master.

12. Jenny L. Davis, "Authenticity, Digital Media, and Person Identity Verification," in *Identities in Everyday Life*, ed. Jan E. Stets and Richard T. Serpe, 93–111 (Oxford: Oxford University Press, 2019).

13. Noah J. Springer, "Publics and Counterpublics on the Front Page of the Internet: The Cultural Practices, Technological Affordances, Hybrid Economies, and Politics of Reddit's Public Sphere" (PhD diss., University of Colorado, Boulder, 2015); Malte Ziewitz, "Evaluation as Governance: The Practical Politics of Reviewing, Rating and Ranking on the Web" (PhD diss., University of Oxford, 2013).

14. Christopher M. Julien, "The Iconic Ghetto, Color-Blind Racism and White Masculinities: A Content and Discourse Analysis of Black Twitter on www.Imgur.com" (MA thesis, University of North Carolina, Greensboro, 2017).

15. Bernward Joerges, "Do Politics Have Artefacts?," *Social Studies of Science* 29, no. 3 (1999): 411–431; Steve Woolgar and Geoff Cooper, "Do Artefacts Have Ambivalence? Moses' Bridges, Winner's Bridges and Other Urban Legends in S&Ts," *Social Studies of Science* 29, no. 3 (1999): 433–449.

16. See Nathan Jurgenson, *The Social Photo: On Photography and Social Media* (London: Verso, 2019).

17. Julia Angwin, Madeleine Varner, and Ariana Tobin, "Facebook Enabled Advertisers to Reach 'Jew Haters,'" *ProPublica*, September 4, 2017, https://www.propublica.org/article/facebook-enabled-advertisers-to-reach-jew-haters.

18. Alex Kantrowitz, "Google Allowed Advertisers to Target People Searching Racist Phrases," *BuzzFeed News*, September 16, 2017, https://www.buzzfeed.com/alexkantrowitz/google-allowed-advertisers-to-target-jewish-parasite-black?utm_term=.px1Y52YxQ#.pqB54857A.

Chapter 5

1. Chelsea Gorrow, "Bystander Arrested as Police Handle Standoff," *Register-Guard* (Eugene, OR), August 22, 2014, http://www.registerguard.com/rg/news/local/32047393-75/woman-arrested-for-crossing-police-line.html.csp.

2. "Jefferson Westside, Eugene, OR Livability," Areavibes, https://www.areavibes.com/eugene-or/jefferson+westside/livability.

3. Jenny L. Davis and James B. Chouinard, "Theorizing Affordances: From Request to Refuse," *Bulletin of Science, Technology & Society* 36, no. 4 (2016): 245

4. Joanna McGrenere and Wayne Ho, "Affordances: Clarifying and Evolving a Concept," in *Proceedings of Graphics Interface 2000: Montreal, Quebec, Canada, 15–17 May 2000* (Montreal: Canadian Human-Computer Communications Society, 2000), 7.

5. Batya Friedman and David G. Hendry, *Value Sensitive Design: Shaping Technology with Moral Imagination* (Cambridge, MA: MIT Press, 2019).

6. Donald A. Norman, *The Design of Everyday Things* (Cambridge, MA: MIT Press, 1998).

7. Gina Neff, Tim Jordan, Joshua McVeigh-Schultz, and Tarleton Gillespie, "Affordances, Technical Agency, and the Politics of Technologies of Cultural Production," *Journal of Broadcasting & Electronic Media* 56, no. 2 (2012): 304.

8. See José Van Dijck, *The Culture of Connectivity: A Critical History of Social Media* (Oxford: Oxford University Press, 2013); Taina Bucher, *If . . . Then: Algorithmic Power and Politics* (New York: Oxford University Press, 2018); Jenny L. Davis, "Curation: A Theoretical Treatment," *Information, Communication & Society* 20, no. 5 (2017): 770–793.

9. Frank Pasquale, *The Black Box Society: The Secret Algorithms That Control Money and Information* (Cambridge, MA: Harvard University Press, 2015); Siva Vaidhyanathan, *Antisocial Media: How Facebook Disconnects Us and Undermines Democracy* (Oxford: Oxford University Press, 2018); Tarleton Gillespie, *Custodians of the Internet: Platforms, Content Moderation, and the Hidden Decisions That Shape Social Media* (New Haven: Yale University Press, 2018).

10. See Michael Oliver and Colin Barnes, *The New Politics of Disablement* (New York: Palgrave Macmillan, 2012).

11. William H. Warren, "Perceiving Affordances: Visual Guidance of Stair Climbing," *Journal of Experimental Psychology: Human Perception and Performance* 10, no. 5 (1984): 683–703.

12. Andrew Kirkpatrick, Joshue O'Connor, Alastair Campbell, and Michael Cooper, "Web Content Accessibility Guidelines (WCAG) 2.1," W3C Recommendation 05 June 2018, World Wide Web Consortium, https://www.w3.org/TR/WCAG21.

13. Stefanie Duguay, Jean Burgess, and Nicolas Suzor, "Queer Women's Experiences of Patchwork Platform Governance on Tinder, Instagram, and Vine," *Convergence: The International Journal of Research into New Media Technology* (June 19, 2018), https://doi.org/10.1177/1354856518781530.

14. Tinder has since made its flagging feature more prominent.

15. Nicole B. Ellison, Charles Steinfield, and Cliff Lampe, "The Benefits of Facebook 'Friends': Social Capital and College Students' Use of Online Social Network Sites," *Journal of Computer-Mediated Communication* 12, no. 4 (2007): 1143–1168; Nicole B. Ellison, Charles Steinfield, and Cliff Lampe, "Connection Strategies: Social Capital Implications of Facebook-Enabled Communication Practices," *New Media & Society* 13, no. 6 (2011): 1873–1892.

16. Hua Wang, Renwen Zhang, and Barry Wellman, "Are Older Adults Networked Individuals? Insights from East Yorkers' Network Structure, Relational Autonomy, and Digital Media Use," *Information, Communication & Society* 21, no. 5 (2018): 681–696; Anabel Quan-Haase, Guang Ying Mo, and Barry Wellman, "Connected Seniors: How Older Adults in East York Exchange Social Support Online and Offline," *Information, Communication & Society* 20, no. 7 (2017): 967–983.

17. Alice Marwick, Claire Fontaine, and danah boyd, "'Nobody Sees It, Nobody Gets Mad': Social Media, Privacy, and Personal Responsibility among Low-SES Youth," *Social Media + Society* 3, no. 2 (2017); Robin Stevens, Stacia Gilliard-Matthews, Jamie Dunaev, Marcus K. Woods, and Bridgette M. Brawner, "The Digital Hood: Social Media Use among Youth in Disadvantaged Neighborhoods," *New Media & Society* 19, no. 6 (2017): 950–967.

18. Saudi Arabia was the last country to maintain legal bans against women drivers.

Chapter 6

1. André Brock, "Life on the Wire: Deconstructing Race on the Internet," *Information, Communication & Society* 12, no. 3 (2009): 344–363; André Brock, "Critical Technocultural Discourse Analysis," *New Media & Society* 20, no. 3 (2018): 1012–1033.

2. Brock, "Critical Technocultural Discourse Analysis," 1016.

3. Brock, "Critical Technocultural Discourse Analysis," 1020 (emphasis in original).

4. Brock, "Critical Technocultural Discourse Analysis," 1023.

5. Sonia Livingstone, "On the Material and the Symbolic: Silverstone's Double Articulation of Research Traditions in New Media Studies," *New Media & Society* 9, no. 1 (2007): 16–24.

6. Ronald E. Day, "Kling and the 'Critical': Social Informatics and Critical Informatics," *Journal of the American Society for Information Science and Technology* 58, no. 4 (2007): 575–582.

7. Lisa Nakamura, "Cultural Difference, Theory and Cyberculture Studies: A Case of Mutual Repulsion," in *Critical Cyberculture Studies*, ed. David Silver and Adrienne Massanari, 29–36 (New York: New York University Press, 2006).

8. Jenny L. Davis, "The End of What People Do Online," *New Criticals* (blog), 2017, http://www.newcriticals.com/the-end-of-what-people-do -online.

9. Roger S. Pressman, *Software Engineering: A Practitioner's Approach* (New York: McGraw-Hill, 2005).

10. Ben Light, Jean Burgess, and Stefanie Duguay, "The Walkthrough Method: An Approach to the Study of Apps," *New Media & Society* 20, no. 3 (2018): 881–900.

11. Michael E. Fagan, "Design and Code Inspections to Reduce Errors in Program Development," *IBM Systems Journal* 38, no. 2/3 (1999): 258.

12. Light, Burgess, and Duguay, "The Walkthrough Method."

13. Susan Leigh Star, "The Ethnography of Infrastructure," *American Behavioral Scientist* 43, no. 3 (1999): 377–391.

14. Light, Burgess, and Duguay, "The Walkthrough Method," 883.

15. See Stefanie Duguay, Jean Burgess, and Nicolas Suzor, "Queer Women's Experiences of Patchwork Platform Governance on Tinder, Instagram, and Vine," *Convergence: The International Journal of Research into New Media Technologies* (June 19, 2018).

16. Light, Burgess, and Duguay, "The Walkthrough Method," 887.

17. Edwin Sayes, "Actor-Network Theory and Methodology: Just What Does It Mean to Say That Nonhumans Have Agency?," *Social Studies of Science* 44, no. 1 (2014): 134–149.

18. Rena Bivens and Amy Adele Hasinoff, "Rape: Is There an App for That? An Empirical Analysis of the Features of Anti-Rape Apps," *Information, Communication & Society* 21, no. 8 (2018): 1051.

19. Bivens and Hasinoff, "Rape: Is There an App for That?"

20. Elizabeth A. Armstrong, Miriam Gleckman-Krut, and Lanora Johnson, "Silence, Power, and Inequality: An Intersectional Approach to Sexual Violence," *Annual Review of Sociology* 44 (2018): 99–122.

21. Bivens and Hasinoff, "Rape: Is There an App for That?," 1052.

22. Bivens and Hasinoff, "Rape: Is There an App for That?," 1053.

23. Batya Friedman and David G. Hendry, *Value Sensitive Design: Shaping Technology with Moral Imagination* (Cambridge, MA: MIT Press, 2019); Batya Friedman, P. Kahn, and Alan Borning, "Value Sensitive Design and Information Systems," in *Human-Computer Interaction in Management Information Systems: Foundations*, ed. Ping Zhang and Dennis F Galletta, 348–372 (New York: Routledge, 2006).

24. Alexei Czeskis et al., "Parenting from the Pocket: Value Tensions and Technical Directions for Secure and Private Parent-Teen Mobile Safety," in *SOUPS '10: Proceedings of the Sixth Symposium on Usable Privacy*

and Security. Redmond, Washington, USA, July 14–16 2010 (New York: ACM, 2010); Mary Flanagan, Daniel C. Howe, and Helen Nissenbaum, "Values at Play: Design Tradeoffs in Socially-Oriented Game Design," in *CHI '05: Proceedings of the SIGCHI Conference on Human Factors in Computing Systems. Portland, Oregon, USA, April 02–07 2005*, 751–760 (New York: ACM, 2005); Batya Friedman and David Hendry, "The Envisioning Cards: A Toolkit for Catalyzing Humanistic and Technical Imaginations," in *CHI '12: Proceedings of the SIGCHI Conference on Human Factors in Computing Systems. Austin, Texas, USA, May 05–10 2012*, 1145–1148 (New York: ACM, 2012); Lisa P. Nathan, Predrag V. Klasnja, and Batya Friedman, "Value Scenarios: A Technique for Envisioning Systemic Effects of New Technologies," in *Proceedings of CHI '07 Extended Abstracts on Human Factors in Computing Systems, San Jose, CA, April 28–May 3, 2007*, 2585–2590 (New York: ACM, 2007)); Katie Shilton, "Values Levers: Building Ethics into Design," *Science, Technology & Human Values* 38, no. 3 (2013): 374–397.

25. Katie Shilton, Jes A. Koepfler, and Kenneth R. Fleischmann, "How to See Values in Social Computing: Methods for Studying Values Dimensions," in *Proceedings of the Seventeenth ACM Conference on Computer Supported Cooperative Work and Social Computing, Baltimore, MD, February 15–19, 2014*, 426–435 (New York: ACM, 2014); Katie Shilton, "Engaging Values Despite Neutrality: Challenges and Approaches to Values Reflection during the Design of Internet Infrastructure," *Science, Technology & Human Values* 43, no. 2 (2018): 247–269.

26. Shilton, "Engaging Values Despite Neutrality."

27. Carl DiSalvo, *Adversarial Design* (Cambridge, MA: MIT Press, 2012).

28. Anthony Dunne and Fiona Raby, *Design Noir: The Secret Life of Electronic Objects* (Berlin: Springer Science & Business Media, 2001).

29. Chantal Mouffe, "Some Reflections on an Agonistic Approach to the Public," in *Making Things Public*, ed. Bruno Latour and Peter Weibel, 804–807 (Cambridge, MA: MIT Press, 2005).

30. DiSalvo, *Adversarial Design*.

31. DiSalvo, *Adversarial Design*, 113

32. Mark Shepard, "Sentient City Survival Kit: Archaeology of the Near Future," in *Proceedings of the Digital Arts and Culture Conference, 2009. After Media: Embodiment and Context, University of California, Irvine, December 12–15, 2009.*

33. See George Herbert Mead, *Mind, Self and Society*, vol. 111 (Chicago: University of Chicago Press, 1934); Jenny L. Davis and Tony P. Love, "Self-in-Self, Mind-in-Mind, Heart-in-Heart: The Future of Role-Taking, Perspective Taking, and Empathy," in *Advances in Group Processes*, ed. Shane R. Thye and Edward J. Lawler, 151–174 (Bingley, UK: Emerald Publishing, 2017); Michael L. Schwalbe, "Role Taking Reconsidered: Linking Competence and Performance to Social Structure," *Journal for the Theory of Social Behaviour* 18, no. 4 (1988): 411–436.

Chapter 7

1. James J. Gibson, *The Senses Considered as Perceptual Systems* (Boston: Houghton Mifflin, 1966), 285.

2. Humanising Machine Intelligence, https://hmi.anu.edu.au.

3. Cambridge Leverhulme Centre for the Future of Intelligence (CFI) (http://lcfi.ac.uk/); Oxford Artificial Intelligence Society (http://oxai.org), Future of Humanity Institute (https://www.fhi.ox.ac.uk), and Centre for Governance of AI (https://www.fhi.ox.ac.uk/govai/); Stanford Institute for Human-Centered Artificial Intelligence (https://hai.stanford.edu); Tsinghua University Institute for Artificial Intelligence (https://gbtimes.com/tsinghua-university-establishes-institute-of-artificial-intelligence); AI Now Institute (https://ainowinstitute.org); DeepMind (https://deepmind.com); OpenAI (https://openai.com); Allen Institute for Artificial Intelligence (https://allenai.org).

4. Laura Robinson et al., "Digital Inequalities and Why They Matter," *Information, Communication & Society* 18, no. 5 (2015): 569–582.

5. André Brock, "Critical Technocultural Discourse Analysis," *New Media & Society* 20, no. 3 (2018): 1012–1030.

6. Philippe Verduyn et al., "Do Social Network Sites Enhance or Undermine Subjective Well-Being? A Critical Review," *Social Issues and Policy Review* 11, no. 1 (2017): 274–302.

7. Amy Orben and Andrew K. Przybylski, "The Association between Adolescent Well-Being and Digital Technology Use," *Nature Human Behaviour* 3, no. 2 (2019): 173–182.

8. Markus Prior, *Post-Broadcast Democracy: How Media Choice Increases Inequality in Political Involvement and Polarizes Elections* (Cambridge: Cambridge University Press, 2006).

9. Manuel Castells, *Communication Power* (Oxford: Oxford University Press, 2013).

10. Jonathan Hopkin and Ben Rosamond, "Post-Truth Politics, Bullshit and Bad Ideas: 'Deficit Fetishism' in the UK," *New Political Economy* 23, no. 6 (2017): 641–655; Stephen Barnard, *Citizens at the Gates* (New York: Palgrave Macmillan, 2018).

11. Maya Indira Ganesh, "Entanglement: Machine Learning and Human Ethics in Driver-Less Car Crashes," APRJA, http://www.aprja.net/entanglement-machine-learning-and-human-ethics-in-driver-less-car-crashes (2017); Robert Sparrow and Mark Howard, "When Human Beings Are Like Drunk Robots: Driverless Vehicles, Ethics, and the Future of Transport," *Transportation Research Part C: Emerging Technologies* 80 (2017): 206–215.

12. Matthew Claudel and Carlo Ratti, "Full Speed Ahead: How the Driverless Car Could Transform Cities," McKinsey & Company (2015), https://www.mckinsey.com/business-functions/sustainability/our-insights/full-speed-ahead-how-the-driverless-car-could-transform-cities.

13. "World's First Gene-Edited Babies Created in China, Claims Scientist," *The Guardian*, November 26, 2018, https://www.theguardian.com/science/2018/nov/26/worlds-first-gene-edited-babies-created-in-china-claims-scientist.

14. Deborah Lupton, "Quantifying the Body: Monitoring and Measuring Health in the Age of mHealth Technologies," *Critical Public Health*

23, no. 4 (2013): 393–403; Deborah Lupton and Gavin J. D. Smith, "'A Much Better Person': The Agential Capacities of Self-Tracking Practices," in *Metric Culture: Ontologies of Self-Tracking Practices*, ed. Btijah Ajana, 57–75 (London: Emerald Publishing, 2018); Gina Neff and Dawn Nafus, *Self-Tracking* (Cambridge, MA: MIT Press, 2016).

15. Gabi Schaffzin, "Reclaiming the Margins in the Face of the Quantified Self," *Review of Disability Studies: An International Journal* 14, no. 2 (2018).

Bibliography

Anderson, Warwick. "From Subjugated Knowledge to Conjugated Subjects: Science and Globalisation, or Postcolonial Studies of Science?" *Postcolonial Studies* 12, no. 4 (2009): 389–400.

Andrejevic, Mark. "Data Collection without Limits: Automated Policing and the Politics of Framelessness." In *Big Data, Crime and Social Control*, edited by Ales Zavrsnik, 111–125. New York: Routledge, 2017.

Angwin, Julia, Madeleine Varner, and Ariana Tobin. "Facebook Enabled Advertisers to Reach 'Jew Haters.'" *ProPublica*, September 4, 2017, https://www.propublica.org/article/facebook-enabled-advertisers-to-reach-jew-haters.

Armstrong, Elizabeth A., Miriam Gleckman-Krut, and Lanora Johnson. "Silence, Power, and Inequality: An Intersectional Approach to Sexual Violence." *Annual Review of Sociology* 44 (2018): 99–122.

Banks, David. "A Brief Summary of Actor Network Theory." *Cyborgology* (blog). *The Society Pages*, 2011. https://thesocietypages.org/cyborgology/2011/12/02/a-brief-summary-of-actor-network-theory.

Barnard, Stephen. *Citizens at the Gates*. New York: Palgrave Macmillan, 2018.

Benjamin, Ruha. "Catching Our Breath: Critical Race STS and the Carceral Imagination." *Engaging Science, Technology, and Society* 2 (2016): 145–156.

Berg, Marc. "The Politics of Technology: On Bringing Social Theory into Technological Design." *Science, Technology & Human Values* 23, no. 4 (1998): 456–490.

Bivens, Rena. "The Gender Binary Will Not Be Deprogrammed: Ten Years of Coding Gender on Facebook." *New Media & Society* 19, no. 6 (2017): 880–898.

Bivens, Rena, and Amy Adele Hasinoff. "Rape: Is There an App for That? An Empirical Analysis of the Features of Anti-Rape Apps." *Information, Communication & Society* 21, no. 8 (2018): 1050–1067.

Bloor, David. "Anti-Latour." *Studies in History and Philosophy of Science Part A* 30, no. 1 (1999): 81–112.

boyd, danah. "Social Network Sites as Networked Publics: Affordances, Dynamics, and Implications." In *A Networked Self*, edited by Zizi Papacharissi, 47–66. New York: Routledge, 2010.

Brayne, Sarah. "Big Data Surveillance: The Case of Policing." *American Sociological Review* 82, no. 5 (2017): 977–1008.

Brock, André. "Critical Technocultural Discourse Analysis." *New Media & Society* 20, no. 3 (2018): 1012–1030.

Brock, André. "From the Blackhand Side: Twitter as a Cultural Conversation." *Journal of Broadcasting & Electronic Media* 56, no. 4 (2012): 529–549.

Brock, André. "Life on the Wire: Deconstructing Race on the Internet." *Information, Communication & Society* 12, no. 3 (2009): 344–363.

Broussard, Meredith. *Artificial Unintelligence: How Computers Misunderstand the World*. Cambridge, MA: MIT Press, 2018.

Bruckner, Matthew Adam. "The Promise and Perils of Algorithmic Lenders' Use of Big Data." *Chicago-Kent Law Review* 93, no. 1 (2018): 3.

Bucher, Taina. *If . . . Then: Algorithmic Power and Politics*. New York: Oxford University Press, 2018.

Burlamaqui, Leonardo, and Andy Dong. "The Use and Misuse of the Concept of Affordance." In *Design Computing and Cognition '14,* edited by John S. Gero and Sean Hanna, 295–311. Basel: Springer, 2015.

Callon, Michel. "The Sociology of an Actor-Network: The Case of the Electric Vehicle." In *Mapping the Dynamics of Science and Technology: Sociology of Science in the Real World*, edited by Michel Callon, John Law, and Arie Rip, 19–34. London: Palgrave Macmillan, 1986.

Callon, Michel, and John Law. "After the Individual in Society: Lessons on Collectivity from Science, Technology and Society." *Canadian Journal of Sociology / Cahiers Canadiens de Sociologie* 22, no. 2 (1997): 165–182.

Castells, Manuel. *Communication Power*. Oxford: Oxford University Press, 2013.

Chemero, Anthony. "An Outline of a Theory of Affordances." *Ecological Psychology* 15, no. 2 (2003): 181–195.

Ciavola, Benjamin T., and John K. Gershenson. "Affordance Theory for Engineering Design." *Research in Engineering Design* 27, no. 3 (2016): 251–263.

Claudel, Matthew, and Carlo Ratti. "Full Speed Ahead: How the Driverless Car Could Transform Cities." McKinsey & Company, 2015. https://www.mckinsey.com/business-functions/sustainability/our-insights/full-speed-ahead-how-the-driverless-car-could-transform-cities.

Cochoy, Franck. "Driving a Shopping Cart from STS to Business, and the Other Way Round: On the Introduction of Shopping Carts in American Grocery Stores (1936–1959)." *Organization* 16, no. 1 (2009): 31–55.

Conole, Grainne, and Martin Dyke. "What Are the Affordances of Information and Communication Technologies?" *ALT-J: Research in Learning Technology* 12, no. 2 (2004): 113–124.

Cooper, Alan, Robert Reimann, and David Cronin. *About Face 3: The Essentials of Interaction Design*. Hoboken, NJ: Wiley, 2007.

Crawford, Kate. "Can an Algorithm Be Agonistic? Ten Scenes from Life in Calculated Publics." *Science, Technology & Human Values* 41, no. 1 (2016): 77–92.

Crawford, Kate, and Tarleton Gillespie. "What Is a Flag For? Social Media Reporting Tools and the Vocabulary of Complaint." *New Media & Society* 18, no. 3 (2016): 410–428.

Crawford, Kate, and Meredith Whittaker. "The AI Now Report: The Social and Economic Implications of Artificial Intelligence Technologies in the Near-Term." AI Now Institute, September 22, 2016. https://ainowinstitute.org/AI_Now_2016_Report.pdf.

Cutting, James E. "Two Ecological Perspectives: Gibson vs. Shaw and Turvey." *American Journal of Psychology* 95, no. 2 (1982): 199–222.

Czeskis, Alexei, Ivayla Dermendjieva, Hussein Yapit, Alan Borning, Batya Friedman, Brian Gill, and Tadayoshi Kohno. "Parenting from the Pocket: Value Tensions and Technical Directions for Secure and Private Parent-Teen Mobile Safety." In *SOUPS '10: Proceedings of the Sixth Symposium on Usable Privacy and Security. Redmond, Washington, USA, July 14–16 2010*. New York: ACM, 2010.

Datta, Amit, Michael Carl Tschantz, and Anupam Datta. "Automated Experiments on Ad Privacy Settings." *Proceedings on Privacy Enhancing Technologies* 2015, no. 1 (2015): 92–112.

Davis, Jenny. "Architecture of the Personal Interactive Homepage: Constructing the Self through Myspace." *New Media & Society* 12, no. 7 (2010): 1103–1109.

Davis, Jenny L. "Authenticity, Digital Media, and Person Identity Verification." In *Identities in Everyday Life*, edited by Jan E. Stets and Richard T. Serpe, 93–111. Oxford: Oxford University Press, 2019.

Davis, Jenny L. "Curation: A Theoretical Treatment." *Information, Communication & Society* 20, no. 5 (2017): 770–783.

Davis, Jenny L. "The End of What People Do Online." *New Criticals* (blog), 2017. http://www.newcriticals.com/the-end-of-what-people-do-online.

Davis, Jenny L. "Triangulating the Self: Identity Processes in a Connected Era." *Symbolic Interaction* 37, no. 4 (2014): 500–523.

Davis, Jenny L., and James B. Chouinard. "Theorizing Affordances: From Request to Refuse." *Bulletin of Science, Technology & Society* 36, no. 4 (2016): 241–248.

Davis, Jenny L., and Tony P. Love. "Self-in-Self, Mind-in-Mind, Heart-in-Heart: The Future of Role-Taking, Perspective Taking, and Empathy." In *Advances in Group Processes*, edited by Shane R. Thye and Edward J. Lawler, 151–174. Bingley, UK: Emerald Publishing, 2017.

Day, Ronald E. "Kling and the "Critical": Social Informatics and Critical Informatics." *Journal of the American Society for Information Science and Technology* 58, no. 4 (2007): 575–582.

Deleuze, Gilles, and Felix Guattari. *A Thousand Plateaus.* Translated by Brian Massumi. London: Continuum, 1987.

Dinsmore, Brooke. "Contested Affordances: Teachers and Students Negotiating the Classroom Integration of Mobile Technology." *Information, Communication & Society* 22, no. 5 (2019): 664–677.

DiSalvo, Carl. *Adversarial Design.* Cambridge, MA: MIT Press, 2012.

Dotov, Dobromir G., Lin Nie, and Matthieu M. De Wit. "Understanding Affordances: History and Contemporary Development of Gibson's Central Concept." *Avant: The Journal of the Philosophical-Interdisciplinary Vanguard* 3, no. 2 (2012): 28–39.

Dourish, Paul. *The Stuff of Bits: An Essay on the Materialities of Information.* Cambridge, MA: MIT Press, 2017.

Duguay, Stefanie, Jean Burgess, and Nicolas Suzor. "Queer Women's Experiences of Patchwork Platform Governance on Tinder, Instagram, and Vine." *Convergence: The International Journal of Research into New Media Technologies* (June 19, 2018). https://doi.org/10.1177/1354856518781530.

Dunne, Anthony, and Fiona Raby. *Design Noir: The Secret Life of Electronic Objects.* Berlin: Springer Science & Business Media, 2001.

Dyer, Richard. *White: Essays on Race and Culture.* 2nd ed. London: Routledge, 2017.

Ellison, Nicole B., Charles Steinfield, and Cliff Lampe. "The Benefits of Facebook 'Friends': Social Capital and College Students' Use of Online Social Network Sites." *Journal of Computer-Mediated Communication* 12, no. 4 (2007): 1143–1168.

Ellison, Nicole B., Charles Steinfield, and Cliff Lampe. "Connection Strategies: Social Capital Implications of Facebook-Enabled Communication Practices." *New Media & Society* 13, no. 6 (2011): 873–892.

Eubanks, Virginia. *Automating Inequality: How High-Tech Tools Profile, Police, and Punish the Poor.* New York: St. Martin's Press, 2018.

Evans, Sandra K., Katy E. Pearce, Jessica Vitak, and Jeffrey W. Treem. "Explicating Affordances: A Conceptual Framework for Understanding Affordances in Communication Research." *Journal of Computer-Mediated Communication* 22, no. 1 (2017): 35–52.

Fagan, Michael E. "Design and Code Inspections to Reduce Errors in Program Development." *IBM Systems Journal* 38, no. 2/3 (1999): 258.

Faraj, Samer, and Bijan Azad. "The Materiality of Technology: An Affordance Perspective." In *Materiality and Organizing: Social Interaction in a Technological World*, edited by Paul M. Leonardi, Bonnie A. Nardi, and Jannis Kallinikos, 237–258. Oxford: Oxford University Press, 2012.

Flanagan, Mary, Daniel C. Howe, and Helen Nissenbaum. "Values at Play: Design Tradeoffs in Socially-Oriented Game Design." In *CHI '05: Proceedings of the SIGCHI Conference on Human Factors in Computing Systems. Portland, Oregon, USA, April 02–07 2005*, 751–760. New York: ACM, 2005.

Flanagan, Mary, and Helen Nissenbaum. *Values at Play in Digital Games.* Cambridge, MA: MIT Press, 2014.

Fortun, Kim. "From Latour to Late Industrialism." *Journal of Ethnographic Theory* 4, no. 1 (2014): 309–329.

Friedman, Batya. "Value-Sensitive Design." *Interactions* 3, no. 6 (1996): 16–23.

Friedman, Batya, and David Hendry. "The Envisioning Cards: A Toolkit for Catalyzing Humanistic and Technical Imaginations." In *CHI '12: Proceedings of the SIGCHI Conference on Human Factors in Computing Systems. Austin, Texas, USA, May 05–10 2012*, 1145–1148. New York: ACM, 2012.

Friedman, Batya, and David G. Hendry. *Value Sensitive Design: Shaping Technology with Moral Imagination*. Cambridge, MA: MIT Press, 2019.

Friedman, Batya, P. Kahn, and Alan Borning. "Value Sensitive Design and Information Systems." In *Human-Computer Interaction in Management Information Systems: Foundations*, edited by Ping Zhang and Dennis F. Galletta, 348–372. New York: Routledge, 2006.

Friedman, Batya, Peter H. Kahn, Alan Borning, and Alina Huldtgren. "Value Sensitive Design and Information Systems." In *Early Engagement and New Technologies: Opening Up the Laboratory*, edited by Neelke Doorn, Dean Schuurbiers, Ibo van de Poel, and Michael E. Gorman, 55–95. Dordrecht: Springer, 2013.

Ganesh, Maya Indira. "Entanglement: Machine Learning and Human Ethics in Driver-Less Car Crashes." *APRJA* (2017). http://www.aprja.net/entanglement-machine-learning-and-human-ethics-in-driver-less-car-crashes.

Gaver, William W. "Situating Action II: Affordances for Interaction: The Social Is Material for Design." *Ecological Psychology* 8, no. 2 (1996): 111–129.

Gibbs, Jennifer L., Nik Ahmad Rozaidi, and Julia Eisenberg. "Overcoming the 'Ideology of Openness': Probing the Affordances of Social Media for Organizational Knowledge Sharing." *Journal of Computer-Mediated Communication* 19, no. 1 (2013): 102–120.

Gibson, James J. *The Ecological Approach to Visual Perception*. Boston: Houghton Mifflin, 1979.

Gibson, James J. *The Ecological Approach to Visual Perception: Classic Edition*. New York: Psychology Press, 2014.

Gibson, James J. *The Senses Considered as Perceptual Systems*. Boston: Houghton Mifflin, 1966.

Gillespie, Tarleton. *Custodians of the Internet: Platforms, Content Moderation, and the Hidden Decisions That Shape Social Media*. New Haven, CT: Yale University Press, 2018.

Gillespie, Tarleton. "The Politics of 'Platforms.'" *New Media & Society* 12, no. 3 (2010): 347–364.

Goldman, Sylvan N. "Commodity Accommodation and Vending Rack." *Official Gazette of the United States Patent and Trademark Office*, vol. 493, August 2, 1938, p. 106. Google Patents, 1938.

Gorrow, Chelsea. "Bystander Arrested as Police Handle Standoff." *Register-Guard* (Eugene, OR), August 22, 2014. http://www.registerguard .com/rg/news/local/32047393-75/woman-arrested-for-crossing-police-line.html.csp.

Granville, Kevin. "Facebook and Cambridge Analytica: What You Need to Know as Fallout Widens." *New York Times*, March 19, 2018.

Harding, S., I. Grewal, C. Kaplan, and R. Wiegman. *Sciences from Below: Feminisms, Postcolonialities, and Modernities*. Durham, NC: Duke University Press, 2008.

Hopkin, Jonathan, and Ben Rosamond. "Post-Truth Politics, Bullshit and Bad Ideas: 'Deficit Fetishism' in the UK." *New Political Economy* 23, no. 6 (2017): 641–655.

Hu, Jun, and George M. Fadel. "Categorizing Affordances for Product Design." Paper presented at ASME 2012 International Design Engineering Technical Conferences and Computers and Information in Engineering Conference, August 2012.

Humphreys, Lee, Veronika Karnowski, and Thilo von Pape. "Smartphones as Metamedia: A Framework for Identifying the Niches Structuring Smartphone Use." *International Journal of Communication* 12 (2018): 2793–2809.

Hutchby, Ian. "Technologies, Texts and Affordances." *Sociology* 35, no. 2 (2001): 441–456.

Ihde, Don. *Technology and the Lifeworld: From Garden to Earth*. Blooming-ton: Indiana University Press, 1990.

Ingold, Tim. "Back to the Future with the Theory of Affordances." *HAU: Journal of Ethnographic Theory* 8, no. 1–2 (2018): 39–44.

Ingold, Tim. *Being Alive: Essays on Movement, Knowledge and Description*. London: Routledge, 2011.

Ingold, Tim. "Culture and the Perception of the Environment." In *Bush Base, Forest Farm: Culture, Environment, and Development*, edited by Elisa-beth Croll and David Parkin, 51–68. London: Routledge, 2002.

Introna, Lucas D. "The Enframing of Code: Agency, Originality and the Plagiarist." *Theory, Culture & Society* 28, no. 6 (2011): 113–141.

Introna, Lucas, and David Wood. "Picturing Algorithmic Surveillance: The Politics of Facial Recognition Systems." *Surveillance & Society* 2, no. 2/3 (2002).

Joerges, Bernward. "Do Politics Have Artefacts?" *Social Studies of Science* 29, no. 3 (1999): 411–431.

Jones, Keith S. "What Is an Affordance?" *Ecological Psychology* 15, no. 2 (2003): 107–114.

Julien, Christopher M. "The Iconic Ghetto, Color-Blind Racism and White Masculinities: A Content and Discourse Analysis of Black Twitter on www.Imgur.com." MA thesis, University of North Carolina, Greens-boro, 2017.

Jurgenson, Nathan. *The Social Photo: On Photography and Social Media*. London: Verso, 2019.

Kahl, Russell, ed. *Selected Writings of Hermann Von Helmholtz*. Middle-town, CT: Wesleyan University Press, 1971.

Kantrowitz, Alex. "Google Allowed Advertisers to Target People Search-ing Racist Phrases." *BuzzFeed News*, September 16, 2017. https://www.buzzfeed.com/alexkantrowitz/google-allowed-advertisers-to-target-jewish-parasite-black?utm_term=.px1Y52YxQ#.pqB54857A.

Katz, Elihu, and Paul F. Lazarsfeld. *Personal Influence: The Part Played by People in the Flow of Mass Communications*. New York: Free Press, 1966.

Kirkpatrick, Andrew, Joshue O Connor, Alastair Campbell, and Michael Cooper. "Web Content Accessibility Guidelines (WCAG) 2.1." W3C Recommendation 05 June 2018, World Wide Web Consortium, 2017. https://www.w3.org/TR/WCAG21.

Koffka, Kurt. *Principles of Gestalt Psychology*. New York: Harcourt, 1935.

Latour, Bruno. *Reassembling the Social: An Introduction to Actor-Network Theory*. New York: Oxford University Press, 2005.

Laurillard, Diana, Matthew Stratfold, Rose Luckin, Lydia Plowman, and Josie Taylor. "Affordances for Learning in a Non-Linear Narrative Medium." *Journal of Interactive Media in Education* no. 2 (2000): 1–19.

Law, John. "Actor Network Theory and Material Semiotics." In *The New Blackwell Companion to Social Theory*, edited by Brian S. Turner, 141–158. Oxford: Wiley-Blackwell, 2009.

Lenander, Aage. "Coin-Operated Lock for a Trolley System Including Especially Shopping and Luggage Trolleys." *Official Gazette of the United States Patent and Trademark Office*, vol. 1047, issue 1, October 2, 1984, p. 142. Google Patents, 1984.

Leonardi, Paul M. "Theoretical Foundations for the Study of Sociomateriality." *Information and Organization* 23, no. 2 (2013): 59–76.

Leonardi, Paul M. "When Flexible Routines Meet Flexible Technologies: Affordance, Constraint, and the Imbrication of Human and Material Agencies." *MIS Quarterly* 35, no. 1 (2011): 147–167.

Lewin, Kurt. *A Dynamic Theory of Personality: Selected Papers*. Translated by Donald K. Adams and Karl E. Zener. New York: McGraw-Hill, 1935.

Light, Ben, Jean Burgess, and Stefanie Duguay. "The Walkthrough Method: An Approach to the Study of Apps." *New Media & Society* 20, no. 3 (2018): 881–900.

Livingstone, Sonia. "On the Material and the Symbolic: Silverstone's Double Articulation of Research Traditions in New Media Studies." *New Media & Society* 9, no. 1 (2007): 16–24.

Lupton, Deborah. "Apps as Artefacts: Towards a Critical Perspective on Mobile Health and Medical Apps." *Societies* 4, no. 4 (2014): 606–622.

Lupton, Deborah. "Quantifying the Body: Monitoring and Measuring Health in the Age of mHealth Technologies." *Critical Public Health* 23, no. 4 (2013): 393–403.

Lupton, Deborah, and Gavin J. D. Smith. "'A Much Better Person': The Agential Capacities of Self-Tracking Practices." In *Metric Culture: Ontologies of Self-Tracking Practices*, edited by Btihaj Ajana, 57–75. London: Emerald Publishing, 2018.

Mace, William M. "James J. Gibson's Ecological Approach: Perceiving What Exists." *Ethics and the Environment* 10, no. 2 (2005): 195–216.

Maier, Jonathan R. A., and Georges M. Fadel. "Affordance Based Design: A Relational Theory for Design." *Research in Engineering Design* 20, no. 1 (2009): 13–27.

Maier, Jonathan R. A., and Georges M. Fadel. "Affordance-Based Design Methods for Innovative Design, Redesign and Reverse Engineering." *Research in Engineering Design* 20, no. 4 (2009): 225.

Maier, Jonathan R. A., and Georges M. Fadel. "Affordance: The Fundamental Concept in Engineering Design." Paper no. DETC2001/ DTM-2170 presented at the ASME Design Theory and Methodology Conference, Pittsburgh, PA, 2001.

Majchrzak, Ann, Samer Faraj, Gerald C. Kane, and Bijan Azad. "The Contradictory Influence of Social Media Affordances on Online Communal Knowledge Sharing." *Journal of Computer-Mediated Communication* 19, no. 1 (2013): 38–55.

Mannell, Kate. "A Typology of Mobile Messaging's Disconnective Affordances." *Mobile Media & Communication* 7, no. 1 (2019): 76–93.

Marshall, Patrick. "Algorithms Can Mask Biases in Hiring." *Sage Business Researcher*, February 15, 2016.

Marwick, Alice E., and danah boyd. "I Tweet Honestly, I Tweet Passionately: Twitter Users, Context Collapse, and the Imagined Audience." *New Media & Society* 13, no. 1 (2011): 114–133.

Marwick, Alice, Claire Fontaine, and danah boyd. "'Nobody Sees It, Nobody Gets Mad': Social Media, Privacy, and Personal Responsibility among Low-SES Youth." *Social Media + Society* 3, no. 2 (2017).

Massanari, Adrienne. "# Gamergate and the Fappening: How Reddit's Algorithm, Governance, and Culture Support Toxic Technocultures." *New Media & Society* 19, no. 3 (2017): 329–346.

Mata, Ivan, Georges Fadel, and Gregory Mocko. "Toward Automating Affordance-Based Design." *AI EDAM: Artificial Intelligence for Engineering Design, Analysis and Manufacturing* 29, no. 3 (2015): 297–305.

McArthur, Leslie Z., and Reuben M. Baron. "Toward an Ecological Theory of Social Perception." *Psychological Review* 90, no. 3 (1983): 215.

McGrenere, Joanna, and Wayne Ho. "Affordances: Clarifying and Evolving a Concept." In *Proceedings of Graphics Interface 2000: Montreal, Quebec, Canada, 15–17 May 2000*. Montreal: Canadian Human-Computer Communications Society, 2000.

McLuhan, Marshall. *Understanding Media: The Extensions of Man*. New York: McGraw-Hill, 1964.

Mead, George Herbert. *Mind, Self and Society*. Vol. 111. Chicago: University of Chicago Press, 1934.

Michaels, Claire F. "Affordances: Four Points of Debate." *Ecological Psychology* 15, no. 2 (2003): 135–148.

Michaels, Claire F., and Claudia Carello. *Direct Perception*. Englewood Cliffs, NJ: Prentice-Hall, 1981.

Mouffe, Chantal. "Some Reflections on an Agonistic Approach to the Public." In *Making Things Public*, edited by Bruno Latour and Peter Weibel, 804–807. Cambridge, MA: MIT Press, 2005.

Mueller, Jennifer C. "Racial Ideology or Racial Ignorance? An Alternative Theory of Racial Cognition." Open Science Framework, March 2, 2019. https://osf.io/fw23k.

Müller, Martin. "Assemblages and Actor-Networks: Rethinking Socio-Material Power, Politics and Space." *Geography Compass* 9, no. 1 (2015): 27–41.

Nagy, Peter, and Gina Neff. "Imagined Affordance: Reconstructing a Keyword for Communication Theory." *Social Media + Society* 1, no. 2 (2015).

Nakamura, Lisa. "Cultural Difference, Theory and Cyberculture Studies: A Case of Mutual Repulsion." In *Critical Cyberculture Studies*, edited by David Silver and Adrienne Massanari, 29–36. New York: New York University Press, 2006.

Nathan, Lisa P., Predrag V. Klasnja, and Batya Friedman. "Value Scenarios: A Technique for Envisioning Systemic Effects of New Technologies." *Proceedings of CHI '07: Extended Abstracts on Human Factors in Computing Systems. San Jose, CA, USA, April 28–May 03, 2007*, 2585–2590. New York: ACM, 2007.

Neff, Gina, Tim Jordan, Joshua McVeigh-Schultz, and Tarleton Gillespie. "Affordances, Technical Agency, and the Politics of Technologies of Cultural Production." *Journal of Broadcasting & Electronic Media* 56, no. 2 (2012): 299–313.

Neff, Gina, and Dawn Nafus. *Self-Tracking*. Cambridge, MA: MIT Press, 2016.

Noble, Safiya Umoja. *Algorithms of Oppression: How Search Engines Reinforce Racism*. New York: NYU Press, 2018.

Norman, Donald A. *The Design of Everyday Things*. Cambridge, MA: MIT Press, 1998.

Norman, Donald A. *The Psychology of Everyday Things*. New York: Basic Books, 1988.

Norman, Donald A. "The Way I See It: Signifiers, Not Affordances." *Interactions* 15, no. 6 (2008): 18–19.

Oliver, Martin. "The Problem with Affordance." *E-Learning and Digital Media* 2, no. 4 (2005): 402–413.

Oliver, Michael, and Colin Barnes. *The New Politics of Disablement.* London: Palgrave Macmillan, 2012.

O'Neil, Cathy. *Weapons of Math Destruction: How Big Data Increases Inequality and Threatens Democracy.* New York: Broadway Books, 2016.

Orben, Amy, and Andrew K. Przybylski. "The Association between Adolescent Well-Being and Digital Technology Use." *Nature Human Behaviour* 3, no. 2 (2019): 173–182.

Parchoma, Gale. "The Contested Ontology of Affordances: Implications for Researching Technological Affordances for Collaborative Knowledge Production." *Computers in Human Behavior* 37, no. Supp. C (2014): 360–368.

Pasquale, Frank. "Algorithms Can Be a Digital Star Chamber." *Aeon*, no. 8. (2015). https://aeon.co/essays/judge-jury-and-executioner-the-unaccountable-algorithm.

Pasquale, Frank. *The Black Box Society: The Secret Algorithms That Control Money and Information.* Cambridge, MA: Harvard University Press, 2015.

Pea, Roy D. "Practices of Distributed Intelligence and Designs for Education." *Distributed Cognitions: Psychological and Educational Considerations* 11 (1993): 47–87.

Pfaffenberger, Bryan. "Social Anthropology of Technology." *Annual Review of Anthropology* 21, no. 1 (1992): 491–516.

Pooley, Jefferson D. "Open Media Scholarship: The Case for Open Access in Media Studies." *International Journal of Communication* 10, no. 2016 (2016): 6148–6164.

Pooley, Jefferson, and Elihu Katz. "Further Notes on Why American Sociology Abandoned Mass Communication Research." *Journal of Communication* 58, no. 4 (2008): 767–786.

Pressman, Roger S. *Software Engineering: A Practitioner's Approach.* New York: McGraw-Hill, 2005.

Prior, Markus. *Post-Broadcast Democracy: How Media Choice Increases Inequality in Political Involvement and Polarizes Elections.* Cambridge: Cambridge University Press, 2006.

Quan-Haase, Anabel, Guang Ying Mo, and Barry Wellman. "Connected Seniors: How Older Adults in East York Exchange Social Support Online and Offline." *Information, Communication & Society* 20, no. 7 (2017): 967–983.

Raynes-Goldie, and Kate Sarah. "Privacy in the Age of Facebook: Discourse, Architecture, Consequences." PhD diss., Curtin University, 2012.

Reed, Edward S. "The Affordances of the Animate Environment: Social Science from the Ecological Point of View." In *What Is an Animal?*, edited by Tim Ingold, 110–126. Abingdon, UK: Routledge, 1988.

Restivo, Sal. "Review Essays: Politics of Latour." *Organization & Environment* 18, no. 1 (2005): 111–115.

Rheeder, Frederik R. L., and Deon Dixon. "Trolley Locking Device." *Official Gazette of the United States Patent and Trademark Office*, vol. 1064, issue 1, March 4, 1986. Google Patents, 1986.

Ricouard, Jacques, and Claude Chappoux. "Coin Lock Device for Shopping Trolleys." *Official Gazette of the United States Patent and Trademark Office*, vol. 1074, issue 3, January 20, 1987, p. 1449. Google Patents, 1987.

Ritzer, George, Paul Dean, and Nathan Jurgenson. "The Coming of Age of the Prosumer." *American Behavioral Scientist* 56, no. 4 (2012): 379–398.

Ritzer, George, and Nathan Jurgenson. "Production, Consumption, Prosumption: The Nature of Capitalism in the Age of the Digital 'Prosumer.'" *Journal of Consumer Culture* 10, no. 1 (2010): 13–36.

Robinson, Laura, Shelia R. Cotten, Hiroshi Ono, Anabel Quan-Haase, Gustavo Mesch, Wenhong Chen, Jeremy Schulz, Timothy M. Hale, and Michael J. Stern. "Digital Inequalities and Why They Matter." *Information, Communication & Society* 18, no. 5 (2015): 569–582.

Savić, Selena, and Gordan Savičić. "Unpleasant Design: Designing out Unwanted Behaviour." In *A Matter of Design: Making Society through Science and Technology*. Proceedings of the Fifth STS Italia Conference. Rome: Società Italiana di Studi sulla Scienza e la Tecnologica, 2014.

Sayes, Edwin. "Actor-Network Theory and Methodology: Just What Does It Mean to Say That Nonhumans Have Agency?" *Social Studies of Science* 44, no. 1 (2014): 134–149.

Scarantino, Andrea. "Affordances Explained." *Philosophy of Science* 70, no. 5 (2003): 949–961.

Schaffzin, Gabi. "Reclaiming the Margins in the Face of the Quantified Self." *Review of Disability Studies: An International Journal* 14, no. 2 (2018).

Schmidt, Richard C. "Scaffolds for Social Meaning." *Ecological Psychology* 19, no. 2 (2007): 137–151.

Schraube, Ernst. "Technology as Materialized Action and Its Ambivalences." *Theory & Psychology* 19, no. 2 (2009): 296–312.

Schrock, Andrew Richard. "Communicative Affordances of Mobile Media: Portability, Availability, Locatability, and Multimediality." *International Journal of Communication* 9 (2015): 1229–1246.

Schwalbe, Michael L. "Role Taking Reconsidered: Linking Competence and Performance to Social Structure." *Journal for the Theory of Social Behaviour* 18, no. 4 (1988).

Senft, Theresa M. "Microcelebrity and the Branded Self." In *A Companion to New Media Dynamics*, edited by John A. M. Hartley, Jean Burgess, and Axel Bruns, 346–354. Chichester, UK: Blackwell, 2013.

Shaw, Robert. "Ecological Psychology: The Consequence of a Commitment to Realism." In *Cognition and the Symbolic Processes*, edited by Walter B. Weimer and David S. Palermo, 159–226. Hillsdale, NJ: Lawrence Erlbaum, 1982.

Shell, Ellen Ruppel. *Cheap: The High Cost of Discount Culture*. New York: Penguin, 2009.

Shepard, Mark. "Sentient City Survival Kit: Archaeology of the Near Future." In *Proceedings of the Digital Arts and Culture Conference, 2009. After Media: Embodiment and Context.* University of California, Irvine, December 12–15, 2009.

Shilton, Katie. "Engaging Values Despite Neutrality: Challenges and Approaches to Values Reflection During the Design of Internet Infrastructure." *Science, Technology, & Human Values* 43, no. 2 (2018): 247–269.

Shilton, Katie. "Values Levers: Building Ethics into Design." *Science, Technology, & Human Values* 38, no. 3 (2013): 374–397.

Shilton, Katie, Jes A. Koepfler, and Kenneth R. Fleischmann. "How to See Values in Social Computing: Methods for Studying Values Dimensions." In *Proceedings of the Seventeenth ACM Conference on Computer-Supported Cooperative Work and Social Computing. Baltimore, MD, 15–19 February 2014*, 426–435. New York, ACM, 2014.

Sparrow, Robert, and Mark Howard. "When Human Beings Are Like Drunk Robots: Driverless Vehicles, Ethics, and the Future of Transport." *Transportation Research Part C: Emerging Technologies* 80 (2017): 206–215.

Springer, Noah Jerome. "Publics and Counterpublics on the Front Page of the Internet: The Cultural Practices, Technological Affordances, Hybrid Economics and Politics of Reddit's Public Sphere." Order No. 3721893, University of Colorado at Boulder, 2015. https://search.proquest.com/docview/1719155030?accountid=8330.

Star, Susan Leigh. "The Ethnography of Infrastructure." *American Behavioral Scientist* 43, no. 3 (1999): 377–391.

Stevens, Robin, Stacia Gilliard-Matthews, Jamie Dunaev, Marcus K. Woods, and Bridgette M. Brawner. "The Digital Hood: Social Media Use among Youth in Disadvantaged Neighborhoods." *New Media & Society* 19, no. 6 (2017): 950–967.

Stoffregen, Thomas A. "Affordances as Properties of the Animal-Environment System." *Ecological Psychology* 15, no. 2 (2003): 115–134.

Suthers, Daniel D. "Technology Affordances for Intersubjective Meaning Making: A Research Agenda for CSCL." *International Journal of Computer-Supported Collaborative Learning* 1, no. 3 (2006): 315–337.

Toffler, Alvin. *The Third Wave.* New York: Bantam, 1980.

Torenvliet, Gerard. "We Can't Afford It! The Devaluation of a Usability Term." *Interactions* 10, no. 4 (2003): 12–17.

Treem, Jeffrey W., and Paul M. Leonardi. "Social Media Use in Organizations: Exploring the Affordances of Visibility, Editability, Persistence, and Association." *Annals of the International Communication Association* 36, no. 1 (2013): 143–189.

Tufekci, Zeynep. *Twitter and Tear Gas: The Power and Fragility of Networked Protest.* New Haven, CT: Yale University Press, 2017.

Turvey, Michael T. "Affordances and Prospective Control: An Outline of the Ontology." *Ecological Psychology* 4, no. 3 (1992): 173–187.

Vaidhyanathan, Siva. *Antisocial Media: How Facebook Disconnects Us and Undermines Democracy.* Oxford: Oxford University Press, 2018.

Van der Hoven, Jeroen, and Noemi Manders-Huits. *Value-Sensitive Design.* Hoboken, NJ: Wiley, 2009.

Van Dijck, José. *The Culture of Connectivity: A Critical History of Social Media.* Oxford: Oxford University Press, 2013.

Vannini, Phillip. "Non-Representational Research Methodologies: An Introduction." In *Non-Representational Methodologies: Reenvisioning Research*, edited by Phillip Vannini, 11–28. New York: Routledge, 2015.

Verbeek, Peter-Paul. "Materializing Morality: Design Ethics and Technological Mediation." *Science, Technology & Human Values* 31, no. 3 (2006): 361–380.

Verduyn, Philippe, Oscar Ybarra, Maxime Résibois, John Jonides, and Ethan Kross. "Do Social Network Sites Enhance or Undermine Subjective Well-Being? A Critical Review." *Social Issues and Policy Review* 11, no. 1 (2017): 274–302.

Vyas, Dhaval, Cristina M. Chisalita, and Alan Dix. "Organizational Affordances: A Structuration Theory Approach to Affordances." *Interacting with Computers* 29, no. 2 (2017): 117–131.

Wajcman, Judy. "The Gender Politics of Technology." In *The Oxford Handbook of Contextual Political Analysis*, edited by Robert E. Goodin and Charles Tilly, 707–721. Oxford: Oxford University Press, 2006.

Wajcman, Judy. "Reflections on Gender and Technology Studies: In What State Is the Art?" *Social Studies of Science* 30, no. 3 (2000): 447–464. http://www.jstor.org/stable/285810.

Wang, Hua, Renwen Zhang, and Barry Wellman. "Are Older Adults Networked Individuals? Insights from East Yorkers' Network Structure, Relational Autonomy, and Digital Media Use." *Information, Communication & Society* 21, no. 5 (2018): 681–696.

Wansink, Brian, Koert van Ittersum, and James E. Painter. "Ice Cream Illusions: Bowls, Spoons, and Self-Served Portion Sizes." *American Journal of Preventive Medicine* 31, no. 3 (2006): 240–243.

Warren, William H. "Perceiving Affordances: Visual Guidance of Stair Climbing." *Journal of Experimental Psychology: Human Perception and Performance* 10, no. 5 (1984): 683–703.

Wicentowski, Danny. "Defying Police 'Do Not Cross' Line Would Be a Crime under Proposed Missouri Bill." *River Front Times* (blog), February 23, 2016. https://www.riverfronttimes.com/newsblog/2016/12/23/defying-police-do-not-cross-line-would-be-a-crime-under-proposed-missouri-bill.

Williams, Apryl A., Zaida Bryant, and Christopher Carvell. "Uncompensated Emotional Labor, Racial Battle Fatigue, and (in) Civility in Digital Spaces." *Sociology Compass* 13, no. 2 (2019): e12658.

Winkler, Till, and Sarah Spiekermann. "Twenty Years of Value Sensitive Design: A Review of Methodological Practices in VSD Projects." *Ethics and Information Technology* (2018): 1–5.

Winner, Langdon. "Do Artifacts Have Politics?" *Daedalus* 109, no. 1 (1980): 121–136.

Winner, Langdon. "Upon Opening the Black Box and Finding It Empty: Social Constructivism and the Philosophy of Technology." *Science, Technology & Human Values* 18, no. 3 (1993): 362–378.

Withagen, Rob, Duarte Araújo, and Harjo J. de Poel. "Inviting Affordances and Agency." *New Ideas in Psychology* 45 (2017): 11–18.

Woolgar, Steve, and Geoff Cooper. "Do Artefacts Have Ambivalence? Moses' Bridges, Winner's Bridges and Other Urban Legends in S&Ts." *Social Studies of Science* 29, no. 3 (1999): 433–449.

"World's First Gene-Edited Babies Created in China, Claims Scientist." *The Guardian*, November 26, 2018. https://www.theguardian.com/science/2018/nov/26/worlds-first-gene-edited-babies-created-in-china-claims-scientist.

Wright, Steve, and Gale Parchoma. "Technologies for Learning? An Actor-Network Theory Critique of 'Affordances' in Research on Mobile Learning." *Research in Learning Technology* 19, no. 3 (2011): 247–258. http://dx.doi.org/10.1080/21567069.2011.624168.

Ziewitz, Malte. "Evaluation as Governance: The Practical Politics of Reviewing, Rating and Ranking on the Web." PhD diss., University of Oxford, 2013.

Index